W9-COZ-247

3 2044 049 136 872

# Domestic Sources of International Environmental Policy

**American and Comparative Environmental Policy**
Sheldon Kamieniecki and Michael E. Kraft, editors

# Domestic Sources of International Environmental Policy
## Industry, Environmentalists, and U.S. Power

Elizabeth R. DeSombre

The MIT Press
Cambridge, Massachusetts
London, England

BOU 5640 - 1/2

GE 170 .D47 2000
DeSombre, Elizabeth R.
Domestic sources of
 international environmental

RECEIVED

AUG 2 0 2001

Kennedy School Library

©2000 Massachusetts Institute of Technology. All rights reserved

No part of this book may be reproduced in any form or by any electronic or
mechanical means (including photocopying, recording, or information storage
and retrieval) without permission in writing from the publisher.

This book was set in Sabon by Crane Composition, Inc. and was printed and
bound in the United States of America
Printed on recycled paper

Library of Congress Cataloging-in-Publication Data

DeSombre, Elizabeth R.
   Domestic sources of international environmental policy : industry, environmentalists, and U.S. power / by Elizabeth R. DeSombre.
   p. cm. — (American and comparative environmental policy)
   Includes bibliographical references and index.
   ISBN 0-262-04179-0 (alk. paper). — ISBN 0-262-54107-6 (pbk. : alk. paper)
   1. Environmental policy—International cooperation. I. Title. II. Series.
GE170.D47   2000
362.7'0526—dc21                                                        99-41745
                                                                           CIP

For Sammy

# Contents

WITHDRAWN

# Series Foreword

Increasingly, international environmental policy is closely intertwined with domestic politics and policy. Action on global climate change is a notable example, where domestic politics greatly shape the positions that nations are prepared to adopt and, thus, the political feasibility of environmental commitments and the pace and effectiveness of policy implementation. Within this context, there are intriguing questions about how new international environmental standards arise and which of the many environmental and natural resource challenges—from protection of the ozone layer to maintenance of the world's biodiversity—rise to the top of the global policy agenda.

In this volume, Elizabeth DeSombre begins with domestic regulation of the environment and asks how such national action affects policy processes at the international level—from agenda setting to policy adoption and enforcement. She argues that, although scholars have frequently documented domestic sources of international policy, we have not systematically studied the conditions under which attempts at internationalization of domestic regulation are made, the means by which such actions are pursued, and the successes of their efforts. To answer those questions, this book analyzes a set of major U.S. environmental regulations within three issue areas and attempts to internationalize them. The three areas are preservation of endangered species, regulation of air quality (including protection of the ozone layer), and conservation of fisheries.

The author finds variation among the three issue areas in the extent and success of attempts at internationalization. She also concludes that the United States is most likely to push to internationalize its domestic environmental regulations when the results are acceptable and advantageous to both environmental and industry actors. In a variation on the theme of politics making for strange bedfellows, DeSombre finds that environmental and industry actors ("Baptists and bootleggers") sometimes find themselves in agreement on desired regulations, albeit for quite different reasons. The success of environmental policy in the international arena reflects a complex mix of factors, including political self-interest, response to economic threats, commitment to environmental or moral standards, and the role played by international policy actors such as epistemic communities.

The study offers new and important insights into the relatively neglected linkages between unilateral and multilateral environmental policy actions. The findings also echo recommendations found in much recent literature about public-private partnerships and the virtues of cooperation between environmentalists and business actors. In exploring these issues, DeSombre suggests how both sets of actors might reconsider their political strategies. She also entices international relations scholars to revisit conventional assessments of selected foreign policy tools such as economic sanctions and to reconsider the bases of political power during an era in which environmental and natural resource issues are gaining in importance and reshaping the relations among nation states.

This book illustrates well the kind of works published in the MIT Press series in American and Comparative Environmental Policy. We encourage books that examine a broad range of environmental policy issues. And we are particularly interested in volumes that incorporate interdisciplinary research and focus on the linkages between public policy and environmental issues, both within the United States and in cross-national settings. We anticipate that future contributions will analyze the policy dimensions of relationships between humans and the environment from either an empirical or a theoretical perspective. At a time when environmental policies are increasingly seen as controversial and new approaches are being implemented widely, the series seeks to assess policy successes and failures, evaluate new institutional arrangements and policy tools, and clarify new directions for environmental policy and politics. These volumes will be written for a wide audience that includes academics, policymakers, environmental scientists and professionals, business and labor leaders, environmental activists, and environmentally minded students. We hope these books will contribute to public understanding of the most important environmental problems, issues, and policies that society now faces and with which it must deal well into the twenty-first century.

Sheldon Kamieniecki, University of Southern California
Michael E. Kraft, University of Wisconsin-Green Bay
Series Editors

# Acknowledgments

This book began its life as a dissertation in the Government Department at Harvard University. During that period my dissertation committee members, Bob Keohane, Lisa Martin, and Gary King, guided me ably through its creation and improvement, as well as through the academic world more broadly. Abe Chayes was a part of the committee until scheduling conflicts precluded his further participation; but during my entire graduate education, and since, he has provided invaluable advice and inspiration.

My graduate school colleagues, Mark Pollack and Shelley Rigger, provided perspective and moral support throughout the process, as did Ron Mitchell, who seems to have had a hand in almost every good thing that has happened to me in my academic career. The same can be said for Audie Klotz, who has been a mentor and friend in matters relating to this project, as well as in everything else.

Several people helped at crucial moments with technical assistance, including Allen Downey, who wrote computer programs to turn impenetrable UN data into usable statistics, and my father, Gene DeSombre, who stepped in to resuscitate failed computer files with alarming frequency.

Thanks are due, as well, to a number of people at governmental, international, and nongovernmental organizations who provided and helped make sense of a wealth of information. Allison Routt at the National Marine Fisheries Service is particularly deserving of mention; she helped to provide and untangle some of the more obscure tuna/dolphin information with spirited good humor.

I have received considerable institutional support throughout the process of researching, writing, and revising this book. I thank the Weatherhead Center for International Affairs (including the Program on Nonviolent

Sanctions, and the International Institutions and Visiting Scholars programs) for support both during graduate school and afterward. Funding in graduate school came as well from the Jacob Javits program and the Mellon Foundation. Colby College is an institution that takes seriously its mission to support scholarship in its faculty; generous funding from both the Social Sciences and the Interdisciplinary Studies divisions enabled me to undertake research crucial to the process of revising and updating this work.

My students at Colby have inspired and challenged me, and a number of them have helped gather some of the information in this book. In particular I thank Betsy Burleson, Kate Litle, Wendy Rice, Amy Rowe, Christina McAlpin, Abby Campbell, and Amy Darling for research assistance and for their sense of adventure. My colleagues in the Government Department and the Environmental Studies Program have provided moral support, as well as evidence that great teaching and impressive scholarship can go hand in hand.

Sammy Barkin has been present at every stage of this project. Without him neither the book nor I would be in as good shape as we currently are. He is my most valued critic and my best friend. Others, including Stephanie LeMenager, Lynda Warwick, Barb Connolly, Michael Ross, David Fairman, Tammi Gutner, and Doug Imig, have provided more support than they perhaps realize. And Tobie, the small furry one, reminded me constantly how important it is to take breaks to play.

Finally, my parents, Nancy and Gene DeSombre, are responsible for guiding my path toward academia. By example they have shown me the value of critical thinking and the joy of education and scholarship. This book is also for them.

# Domestic Sources of International Environmental Policy

# 1

## Introduction and Overview

Until recently, most yellowfin tuna fishers encircled dolphins with purse-seine nets to catch the tuna that schooled below. In the process, dolphins drowned and their numbers worldwide dwindled. Now, many states have banned the use of purse-seine nets, the Inter-American Tropical Tuna Commission (IATTC) has strict regulations to prevent dolphin mortality, and dolphins are thriving. How did such a change in international action happen? In this case, the international regulation began with a U.S. law protecting marine mammals. U.S. tuna fishers, afraid that less efficient dolphin-safe methods of catching yellowfin tuna would put them at a competitive disadvantage compared to foreign tuna suppliers, joined the call of environmentalists for worldwide dolphin protection. Backed by U.S. sanctions against tuna caught in ways that harms dolphins, U.S. parties pushed for multistate acceptance of regulations to protect dolphins. Following tuna embargoes by the world's largest tuna market, several battles in the General Agreement on Tariffs and Trade (GATT), and a growing public sentiment in favor of dolphin protection, even the most recalcitrant states changed their tuna-fishing practices. In addition members of the IATTC, as well as many outside the organization, agreed to international dolphin protection measures.

How do international standards arise? Which international issues will the world choose to address? There are a number of problems that arise in the world that cannot be solved unless multiple states undertake concerted and compatible regulation. Arms control works only if the states in question all agree to limit their acquisition of weapons. Free trade is not actually free unless a number of states remove trade barriers. Instruments of diplomacy are successful only if multiple states adopt them. Ozone depletion cannot be addressed by the actions of one state, no matter how strong, if other states continue to emit ozone-depleting substances.

One important way that regulations in general appear on the international scene is through the internationalization of regulations that one or more states have undertaken domestically. One state that has taken action to regulate its contribution to an international problem attempts to persuade others to do the same. International regulations against slavery began with domestic regulations by Britain, which were then applied to its colonies and later negotiated and enforced internationally.[1] The creation of the Long-Range Transboundary Air Pollution (LRTAP) Agreement followed from initial domestic action to regulate air pollution by Scandinavian countries that showed early awareness of the problem.[2] Obviously not every state is always capable of convincing the international system to adopt regulations it has already undertaken, even a state with considerable international influence. But since, in some situations, individual states have been able to convince others to adopt their domestic regulations, it is useful to study the conditions under which this phenomenon could occur. Ascertaining which domestic environmental regulations are candidates for internationalization, and which ones are likely to be internationalized successfully, can give us insight into the international regulatory process overall. It also can illuminate some of the dynamics behind international cooperation, agenda setting, and enforcement or creation of international norms.

Attempts to address international problems often begin with domestic regulation. One state has an interest in an issue, for whatever reason, and regulates its own behavior. Although such regulation may be illogical to some extent from a commons perspective, domestic pressure can lead to a state's

---

[1] Once international regulation had been passed outlawing transportation of slaves on the high seas, Britain unilaterally worked to enforce these regulations. See Paul E. Lovejoy, *Transformations in Slavery: A History of Slavery in Africa* (Cambridge: Cambridge University Press, 1983), pp. 282–87; David Eltis and James Walvin, eds., *The Abolition of the Atlantic Slave Trade* (Madison, University of Wisconsin Press, 1981).

[2] See, for instance, Marc A. Levy, "European Acid Rain: The Power of Tote-Board Diplomacy," in *Institutions for the Earth: Sources of Effective International Environmental Protection,* ed. Peter M. Haas, Robert O. Keohane, and Marc A. Levy, (Cambridge, Mass.: MIT Press, 1993), p. 17. One of the main reasons for early regulation by Scandinavian countries was that they were the most harmed by the acid precipitation that resulted from transboundary air pollution. In addition, the transboundary nature of the problem meant that they could not solve it solely through domestic regulations.

adoption of domestic regulation on issues that have international components. United States efforts to regulate access to pornography on the Internet will do little to address questionable material on foreign web sites but have strong domestic support. Regulation of corporate tax policy might not influence companies that can transfer assets abroad, but such regulations are nevertheless sometimes first passed on the domestic level. These regulations may confer domestic benefits that are sufficient to create pressure for their passage. They may even have the support of some members of industry who could gain by the regulatory barriers to entry to other firms. For example, regulations that require that all packaging used be taken back by the seller for recycling make foreign firms less able to compete in the market since their costs of complying may be prohibitively high.[3] But these regulations may not be able to address completely the problems that, for some actors at least, they were created to address.

That states with domestic regulations attempt to gain adherence to these regulations by other states is therefore not surprising. But the conditions under which such attempts are made, the means by which they are made, and the success of these attempts have not been systematically investigated as yet.

Environmental problems form an important subset of concerns that cross borders and cannot be addressed sufficiently without international action. Garrett Hardin characterized certain environmental issues as mirroring the "tragedy of the commons,"[4] to point to the difficulty of addressing such problems individually. In his analogy, no individual herder who regulates grazing of cows on common land prevents environmental damage, because others have incentives to continue to graze their cows, and even to increase the size of their herds. It is only in the case of restraint by all herders, unlikely in the anarchy of a commons, that the commons can be saved. Likewise, many environmental issues cannot be addressed by individual states. One state's restrictions on hunting wild birds will do little good if the birds migrate to a state without such restrictions. A decrease in the

---

[3] For other examples of reasons industry may support domestic environmental regulation, see chapter 2, as well as Kenneth A. Oye and James H. Maxwell, "Self-Interest and Environmental Management," *Journal of Theoretical Politics* 6, no. 4, 1995: 607–8.

[4] Garrett Hardin, "The Tragedy of the Commons," *Science* 162, (13 December 1968): 1243–48.

amount of $CO_2$ emitted by one state will do little to prevent global climate change if other states continue to increase their carbon output. For many environmental issues, therefore, individual state actions have little effect unless other states take similar measures.

Some environmental issues are truly local, and can be addressed without resorting to internationalization. Pollution of a local lake, for instance, requires only action by proximate actors. This book deals predominantly with environmental issues with transboundary environmental effects. Domestic regulations to address environmental issues that are purely domestic can nevertheless set in motion a variety of domestic pressures for internationalization. It should also be noted that not all environmental problems are commons problems, in the sense that the polluters bear the same degree of the costs of the pollution. Some pollution issues are purely transboundary, in which "upstream" states export pollution elsewhere. These issues may have a different dynamic than purely commons issues, but they still require action by multiple states to address the environmental problem.

Attempts to internationalize domestic environmental regulations are found across a variety of states with advanced environmental protection policies. The European Union (EU) is responsible for a large number of internationalization attempts, particularly relating to the preservation and humane treatment of animal species. It banned the use of leghold animal traps and banned imports of furs from countries that did not have equivalent regulations.[5] The EU also banned wildlife trade with Indonesia because Indonesia's wildlife protection policies were not equivalent to those of other countries.[6] And it tried diplomatic and then economic pressure to convince Norway and Canada to prohibit the killing of baby seals.[7] Within the EU, as well, individual member states with stronger environmental protection legislation work to convince the EU as a whole to create supranational environmental legislation so that all EU members face the same regulations. Recent efforts to ban driftnet fishing by all EU members are an example of this process.

[5] European Council Regulation 3254/91, 1991 O.J. (L 308).

[6] European Council Regulation 3626/86, "List of Prohibited Species and Countries" 1982 O.J. (L 384) Annex C2.

[7] European Council Directive 83/129, 1981 O.J. (L 91) 30; made permanent by Council Directive 89/370, 1989 O.J. (L 103).

Canada has also worked to internationalize its domestic environmental policies. It pressed for an agreement on acid rain with the United States when it already had in place some domestic legislation to address air pollution.[8] Canada has been particularly active in its attempts to gain adherence to its fisheries regulations by other states, even going so far as to shoot at Spanish fishing vessels that were taking turbot that Canadian fishers were prohibited from taking.[9]

The phenomenon of internationalization of environmental policies is investigated here with respect to the United States. The United States has "among the most stringent environmental regulatory requirements in the world,"[10] and has generally been a leader in international environmental regulation.[11] Beginning a study of internationalization of domestic policy by examining the way in which it happens in the United States has two advantages. First, it holds constant the domestic political process of the state while examining the conditions under which domestic regulations will be internationalized. It would be more difficult to examine those conditions when studying several states with a variety of domestic policy structures and governments. Second, the United States is important in the international environmental policy arena, owing both to its economic size and influence and to its general leadership on environmental issues. Examining the

[8] See, for example, John Fraser, M. P., "The Politics of a Shared Environment," in *Clean Air for North Americans: Acid Rain in Canadian-American Relations,* Report of a Conference Sponsored by the Canadian Studies Program, Columbia University, and issued jointly with The World Environment Center, New York City, June 1981, pp. 45–54. In the case of a U.S.-Canadian Acid Rain agreement, both sides had domestic regulations of different sorts before working to create a bilateral agreement on the issue.

[9] While some aspects of Spain's fishing activity could be seen as noncompliance with Northwest Atlantic Fisheries Organization regulations, division of the quotas under the regulations was nonbinding and the European Union lodged a legal objection to the quota division. Spain contended that Canada was trying to impose its law on European states. See "Canadians Seize Spanish Trawler," *Toronto Star,* 10 March 1995, p. A1 ff.

[10] Richard B. Stewart, "Environmental Regulation and International Competitiveness," *Yale Law Journal* 102 (1993): 2046.

[11] See, for example, Jessica Tuchman Matthews, "Implications for U.S. Policy," in *Preserving the Global Environment: The Challenge of Shared Leadership,* ed. Jessica Tuchman Matthews. (New York: W. W. Norton), pp. 315–16.

internationalization of U.S. domestic policies is therefore likely to give insight into the origins of a large percentage of international environmental regulations. The United States, with a wide range of domestic environmental regulations that exceed international standards, has a large set of domestic regulations that might be candidates for internationalization. And because of the international importance of the United States, it can influence international action, through diplomacy or economic threats. Those who argue that international regulation is determined or initiated by hegemons point to the necessity of the existence of a powerful state for such regulation. The United States is certainly one of the states able to play this role in environmental politics.

It should be noted at this point that there are two possible strategies for examining the relationship between domestic and international environmental regulations, of which this study chooses one: to analyze regulations on the domestic level to discover which ones the state in question attempts to internationalize and which ones, in fact, end up being adopted by other states. The other approach would be to begin with a set of international regulations and examine their origins to ascertain which ones began with domestic regulations. The most simple set to examine would be those contained in international treaties, though doing so would miss the set of convergent domestic environmental regulations that are not codified in treaties but that may have had their origins in the domestic regulations of one state. Although this is a useful way to study the issue, it is not the approach pursued here.

This book therefore begins with a set of major U.S. environmental regulations within three issue areas and determines which of these the United States attempts to convince other states to adopt. Then, within the set of U.S. regulations that the United States attempts to internationalize, it examines which ones other states actually do take on, and why. This type of analysis therefore misses a certain set of international environmental regulations: those that never start on the domestic level. As the international issues discussed here show, however, a significant portion of international environmental regulations do begin on the domestic level. This analysis, therefore, provides an important piece of the puzzle in explaining the emergence of international environmental agreements. In addition, as the fisheries cases show, regulation that begins internationally can still be subject to pressures for further internationalization.

The three issue areas examined are preservation of endangered species, regulation of air quality, and conservation of fisheries. These issues are sufficiently circumscribed so that all major U.S. legislation relating to these issues can be examined to determine the extent to which the United States attempts to internationalize regulations within each issue area. These issue areas present regulations of very different types within the broader issue area of the environment. There is significant variation in the extent and the success of attempts at internationalization, and in the variables with which to test the explanations about likelihood and success of attempted internationalization.

For the purposes of this project, internationalization is defined as an official governmental attempt to gain adherence by other states to a level of environmental regulation similar to that in effect for the state in question on a particular issue.[12] There are two stages in the process. At one stage, a decision is made, within the U.S. policymaking process, about which domestic regulations the United States government will attempt to convince other states to adopt. At this stage the target states in question do not actually have to adopt the standard for an instance to be considered an attempt at internationalization; it is sufficient that the United States makes a governmental effort to convince them to adopt similar regulations. Symbolic action within international environmental politics being prevalent, exhortatory statements about the importance of international regulation are not discussed here. To be considered serious attempts at internationalization, sustained diplomatic efforts must be initiated or explicit threats communicated. In practice, these are generally initiated by congressional legislation. At a second stage the success of the internationalization attempt is determined. To what extent does the United States succeed in convincing other states to adopt its regulatory policies? What determines the success of internationalization attempts?

[12] Note that this definition is markedly different from that used by others who have written about related phenomena under the same term. Robert Keohane and Helen Milner examine "the processes generated by underlying shifts in transactions costs that produce observable flows of goods, services and capital." Robert O. Keohane and Helen V. Milner, *Internalization and Domestic Politics* (Cambridge: Cambridge University Press, 1996), p. 4. Miranda Schreurs and Elizabeth Economy use the term to indicate that environmental problems are increasingly addressed on the international level. Miranda A. Schreurs, and Elizabeth C. Economy, *The Internationalization of Environmental Protection* (Cambridge: Cambridge University Press, 1997). The phenomena examined in both these volumes relate to those examined here, though the specific questions asked are different.

These stages are not isolated from each other. Temporally, Stage I precedes and thereby influences Stage II. Measures cannot be applied to foreign states unless they have been approved politically at the first stage. An analysis of success at Stage II is therefore by definition an analysis of the extent to which policies that succeeded within the domestic political process influence action by the international community; that analysis will not consider all the various types of policies that could have originally been undertaken in an attempt at internationalization. An analysis of success at Stage II therefore examines the extent to which the policies that have already made it past the domestic political process influence action by the international community; it cannot examine all the various types of policies that could have been undertaken in an attempt at internationalization. In this way, the choices made at the first stage influence the possibilities at the second stage.

But influence may work in the other direction, as well. It is not unreasonable to imagine anticipation effects of Stage II at Stage I. Owing to the controversial nature of attempts to influence international policy, Congress may be more willing to undertake internationalization attempts when it foresees success. Actors may push the government hardest to work for internationalization of those regulations they believe foreign states will actually adopt. As chapters 2 through 5 show, however, there *are* some domestic actors who benefit from internationalization attempts, whether or not these attempts are successful. These actors may thus push for internationalization without regard to their likely success. In the case of internationalization of dolphin protection measures, for example, tuna fishers might actually have been better off if attempted internationalization had been unsuccessful. In that case, foreign tuna caught in dolphin-unfriendly ways would have been kept out of the United States and would therefore not have competed with tuna harvested by U.S. fishers. But to the extent there is an anticipation effect, success at Stage II should be even greater than one would expect from a random set of internationalization policies. Any variation in success of internationalization policies should therefore be that much more significant. Regardless, the fact that there is variation in success provides the opportunity to draw conclusions regarding the types of factors, ceteris paribus, that lead to success in influencing international policy.

In both stages of the internationalization process various environmental and economic factors are potentially explanations of how a domestic environmental regulation is pushed and then ultimately adopted internationally. These various explanations rest upon different assumptions that are made about the role the domestic government plays in making policy and the ways in which international cooperation emerges. As such they provide an opportunity to note certain conditions under which these assumptions are likely to hold true.

## Stage I: Sources of Internationalization

The first part of the issue of internationalization of policy is to determine when states will decide to push their domestic regulations internationally. This study therefore first ascertains the conditions under which the United States will attempt to internationalize domestic environmental regulations. Regulations dealing with environmental protection have two possible types of impacts: environmental and economic. Both the environmental and the economic costs and benefits of such regulations can provide incentives for reaction to them. Those who study international environmental action address these factors as possible underlying causes of international environmental cooperation.

Applying this logic not only to international environmental policy but also to the domestic actions that may lead to international regulation yields two types of predictive explanations for the conditions under which we would expect attempts at internationalization. The first is that environmental externalities will drive attempts to encourage other states to adopt environmental regulations similar to those in the United States. When environmental problems cannot be solved without international action, the United States pressures others to adopt regulations to address the issues. Environmental degradation is, almost without fail, the stated reason for pursuing international environmental regulation. Environmental organizations are vocal in their support of expanding environmental protection internationally as a necessary step in mitigating environmental problems.

A second explanation looks to the economic harm suffered by the regulated domestic industries and the potential for economic gain offered by internationalization. When U.S. industry has higher production costs than its international competitors due to regulation, or when some industry has developed a substitute for regulated substances, either type of industry will gain from international adoption of the domestic regulations. According to this second explanation, it is under these conditions that the United States will push for internationalization. Although it is an important element in trade policy, this perspective has largely been ignored in much environmental policy literature.

This study examines these general explanations (environmental and economic), as well as variations of each. The specific argument derived is a combination of the two. The United States typically pushes to internationalize those domestic environmental policies that would be advantageous on the international level for both economic and environmental reasons. When these two incentives coincide, the chances for action are greatest. Without agreement between the two camps—industry and environmentalist actors—the government is not willing to undertake any but the simplest political action to convince other states to address international environmental problems. Domestic regulation may begin with environmental concerns, but industry actors are essential to the process of internationalization. The trade effects of environmental regulations mobilize actors within and across states in ways that the environmental aspects of the policies alone will not. But without environmentalist pressure, as well, to broaden regulation internationally, the United States does not attempt to do so.

The U.S. government decides in favor of internationalization when industry actors find themselves with a similar goal to that of environmentalists: increasing the number of states subject to the type of environmental regulation already imposed by the United States on its domestic actors. Industry actors and environmentalists, in fact, rarely have identical goals. Typically, during the creation of the initial domestic legislation, they are at odds. Once a regulation is in place, moreover, the regulated industry initially works for its removal. But both groups also have, somewhere on their list of priorities, an interest in subjecting other states to similar regulations. Both are more likely to attain this shared goal by working together. The industry actors gain the respectability of the environmentalists who are working for the good of the earth; the environmentalists gain the economic clout and political influence that can be wielded by unified economic interests.

The empirical analysis examines both the environmental and economic implications of broadening environmental regulations. On the economic side, this book examines both those who lose from the domestic environmental regulation and those who benefit from the internationalization (often overlapping but not necessarily identical segments of industry). And while industry involvement is generally necessary for the United States to push internationalization, such efforts are unlikely to go forward without the presence of environmentalists working for the same goal. On the environmental side, the empirical analysis examines which environmental problems cause the most harm, travel most across borders, and gain the most attention of environmental organizations. These problems provide the potential environmental push for internationalization.

Within the three environmental issue areas, environmentalists are most important for internationalizing regulations relating to endangered species. Industry actors are most important for internationalizing regulations relating to fisheries. In no case, however, is one set of actors able to push forward an attempt at internationalizing without engaging the interests of the other group. Even attempts to internationalize endangered species regulations that would seem to have no industry interest have been pursued using legislation passed initially with support by an industry-environmental coalition. Also important to note are the regulations that the United States does not attempt to internationalize. Among them are several that provide either strong environmental *or* economic incentives for internationalization. With only one side pushing for international adoption of these rules, the United States is unwilling to undertake such an effort. Both the economic clout of industry and the moral suasion of environmentalists are required.

### Stage II: Success of Internationalization

The second part of this book considers the effects of U.S. efforts to internationalize domestic environmental regulations. Do other states adopt the environmental regulations in question? What determines how successful U.S. efforts will be? Research in this area attributes success in convincing other states to adopt regulations to a variety of factors about the states involved, as well as about the methods by which internationalization is attempted. These explanations come from different understandings within international relations theory of the determinants of international cooperation. Environmental politics is a frequent focus of studies of international cooperation because of the opportunities for mutual gain from collective action in addressing environmental problems. These studies often view the environment as an arena of low politics where friendly interactions make agreement on cooperative outcomes relatively simple. The analysis here demonstrates, however, that power and threat play a central role in the creation and adoption of international environmental regulation.

States can be expected to adopt common regulations for three different types of reasons. One is that a state will adopt regulations because it thinks doing so is the "right" thing to do regardless of self-interest or fear of consequences. There is a social component to this type of explanation, found also in those who examine legalist sources of international cooperation. These theories examine the legitimacy of the internationalization attempt as indicative of its success. A second reason that a state will adopt a regulation pushed by another state is that it is in its interest to do so. This analysis is similar to liberal arguments for state action that expect international cooperation when states will mutually gain. This approach addresses the inherent advantages to the target states of adopting the regulations in question, something often assumed when examining international environmental problems. A third reason for a state to adopt an internationalized regulation is that it fears it will be harmed by the other state if it does not do so. This explanation also represents self-interest, but of a different character. This perspective is similar to that taken by realist explanations that see power as the driving force in international relations. These explanations rely upon theories that expect international cooperation to arrive only when driven and coerced by the powerful members of the international community. This explanation looks to the characteristics of the sending and the target states that make threats effective. These three explanations of international adoption of regulation are evaluated in the cases examined here.[13]

The specific argument put forth here relies most heavily on a variant of the realist hypothesis, but also relies on the credibility gained from the involvement of the coalition of environmental and economic interests observed at Stage I. States adopt environmental regulations the United States is pushing internationally when they fear retaliation if they do not do so. Economic threats often underpin efforts at internationalization.

---

[13] These categories are similar to those examined by Ian Rowlands, *The Politics of Global Atmospheric Change* (Manchester: Manchester University Press, 1995). His "knowledge" category is covered in part by legitimacy and in part by environmental self-interest as examined here.

The most important aspect in predicting success is the market power that the internationalizing state has over the states it is trying to persuade, relative to the costliness to the target state of adopting the regulation. But, since market power is only a credible threat if the target states believe it will be used, the presence of the domestic actors who actually gain from the imposition of economic restrictions on the target states makes the threat of superior market power credible. As will be explained later, the same domestic coalition that initially worked to pass internationalizing legislation may not be active on the issue in which the legislation is imposed; the presence of the domestic coalition then varies at Stage II as well as at Stage I.

Within the internationalization attempts examined, the states over which the United States has the greatest degree of market power are those that tend to adopt any particular set of internationalized regulations. This tendency indicates that power concerns do play an important role in determining international cooperation. But U.S. power does not always prevail. It is only the particular economic power threatened for inaction in adopting regulations that counts. The relative economic health of the states in question matters much less than the extent to which the United States is an important market for the goods it uses to threaten the target state. Moreover, the evidence from Stage I shows that the form these economic threats will take is determined by the particular coalition of environmental and industry actors needed to get the internationalization legislation passed initially. The United States does not simply threaten with its most powerful economic tools; so the success of economic threats internationally also comes from the domestic action that determines their form.

## Overview and Findings

This study examines the two stages in internationalization to ascertain the conditions under which domestic regulations will be pushed on an international level, and when such attempts will result in the adoption of the regulations in question by the target states.

The first stage concerns the decision by a state to work for internationalization of its domestic environmental regulations. Chapter 2 derives the explanations for internationalization due to environmental and economic externalities and formulates the argument about the relationship of those externalities to each other. Chapters 3 through 5 then evaluate the explanations using the empirical material in the three issue areas: preservation of endangered species, air pollution regulation, and conservation of ocean fishery resources. This first section, as a whole, argues that we should expect to find attempts at internationalization when domestic environmental and economic interests combine to benefit from adoption by other states of the environmental regulations in question. It is when the two interests converge that the U.S. government will undertake serious action to convince other states to adopt regulations similar to those the it has already adopted.

The second stage concerns the extent to which the target states adopt the regulations that the United States is pushing them to adopt. Chapter 6 derives explanations for the conditions under which states are successful at imposing policies on other states, on the basis largely of the advantages of the regulations and the threats that the sending state uses. It presents the argument about the conditions under which these attempts at internationalization are likely to be successful, on the basis of credible threats by the internationalizing state to use market power. Chapter 7 examines the cases from the first section to evaluate the explanations for the conditions under which internationalization attempts will be successful. The second section, as a whole, argues that target states will adopt the regulations pushed by internationalizing states that make a credible threat to impose sanctions in an area in which it has dominant market power. This argument is not in itself surprising to those who see international relations as a realm of power and threats, though it may be to those who see environmental politics as a separate, more friendly, phenomenon. In fact, one of the things this process demonstrates is that environmental politics is like any other kind of politics played on the international scene, despite the perception of its low-politics character and potential for mutual gain. This analysis does, however, show which particular aspects of power will be important in addressing this type of internationalization.

But the story is not that simple. If internationalization requires that the sending state credibly threaten an economic measure whose imposition will actually benefit it, and make that threat to a market sector in which the target state is dependent on the sending state, why should internationalization ever fail? Given that the sending state makes a decision about how to work for internationalization in the process of deciding to do so at all, why does the sending state ever attempt internationalization in a way that would lead to anything but success? Because of the two-stage process of internationalization. The way internationalization is pursued at Stage I limits the options that the sending state can use to pursue its goal. The members of the domestic coalition necessary to pass internationalizing legislation domestically take part in the process for their own reasons. Environmentalists may want to use any means available to persuade target states to adopt environmental regulations, but the means available to them turn out to be those favored by industry actors, without whom the environmentalists seldom have the clout to pass internationalizing legislation. Industry actors push internationalization not only for the possible gains from holding foreign competitors to the same standards they bear, but also for the possibility that if internationalization fails, these competitors will be kept out of the U.S. market. The type of threats they are willing to make are limited to those that focus on the particular resources relating to the regulation in question.

Success varies, then, because the methods of internationalization rarely do. The shape of the domestic legislation determines which domestic actors will be affected enough to want to push for internationalization and in what way. In fact, the very domestic legislation that begins the process of internationalizing can undermine itself if it eliminates the industry coalition partner that would help push the regulation on the international level. The content of the domestic legislation can therefore tell us a lot about the eventual success of the legislation's internationalization.

## Implications

A study about U.S. internationalization of domestic environmental commitments can contribute to an understanding of internationalization processes of environmental regulation by other states, of internationalization processes in issue areas beyond the environment, and of the international regulation process itself, more broadly. It offers comments as well on the regulatory effects of free trade, the origins of international cooperation, the usefulness of economic sanctions, and the importance of considering the interaction between domestic and international politics when analyzing either individually.

There are certainly states other than the United States that pursue internationalization of domestic environmental regulations, and anecdotal evidence suggests that some of these do so in response to pressure from both environmentalists and industry actors. Canada's willingness to take on Spain and the European Union over their refusal to adopt restrictions on their catch of turbot came not only from a general environmental concern for a declining resource, but also from support from the Canadian fishers who did uphold the restrictions. It remains to be investigated to what extent the type of coalition observed within the United States is a necessary or sufficient condition elsewhere for pursuing internationalization. There is evidence, however, that this type of coalition does operate within political systems different from those observed in the United States.

This phenomenon may operate within different issue areas as well. There are concerns other than environmental protection for which moral and economic components may combine in the drive for international policy. Support for international regulations to end prison labor and the codification within the GATT of the acceptance of trade restrictions on goods made from prison labor surely came both from those who opposed maltreating prisoners as well as from those whose industries did not want to compete with the unpaid labor used by others. In the debate over the impacts of free trade on regulatory policies, many participants fear that increasing liberalization of trade policies will negate protections for the environment, for health, for safety, and for workers, since states with greater levels of protection will be less able to compete internationally. They fear that these states either will be tempted to remove such protections or that other states will conduct a "race to the bottom" to lure industry with promises of lower-cost production. The results here show that the opposite phenomenon is possible, as well. Greater international competition gives an incentive to those states with strict domestic regulatory policies to work for greater international protections. Both the regulated industry and the subjects of regulatory protection may thereby benefit.

This study also comments on the process by which international cooperation emerges. It provides evidence of the potentially conflictual nature of the process that produces international regulation. Even in cases studied here where multilateral cooperation eventually emerged, the possibility of economic sanctions should cooperation fail was sometimes either explicit or implicit and occasionally invoked. While it would be wrong to conclude that international cooperation emerges only within a context of threats, the prevalence of these threats within international issues that are generally seen to have been negotiated without much conflict suggests that they play a bigger role in international cooperation than is sometimes acknowledged.

There are also lessons to be drawn from this project about the use and effectiveness of sanctions. Many see sanctions as largely ineffective foreign policy tools. But the sanctions evaluated are often export sanctions, in which sending-state suppliers bear the costs of not being allowed to sell their products abroad and therefore do not support imposition or continuation of sanctions. Without domestic support in the sending-states, target states may not believe that threatened sanctions will be imposed or that imposed sanctions will last long. To the contrary, the sanctions examined here are generally quite credible as threats, since the domestic industries in the sending state actually gain from imposing them (sometimes more than they would if the target state acquiesced and there were no need to impose sanctions). This work, therefore, suggests revisions to the conventional wisdom on the effectiveness of sanctions and offers specific strategies for using sanctions in a way that can be effective.

Finally, this study provides further evidence that domestic politics and international relations are not distinct entities. International action may originate in domestic regulation; even in cases where domestic action alone cannot sufficiently address an international problem, those looking toward the international system may work to regulate domestically first. Thus, domestic actors play an important role in determining the shape of the international regulation sought and the means by which it is pursued. As Robert Putnam points out, international negotiators are playing a "two-level game,"[14] negotiating with both domestic constituencies and foreign counterparts. In the instances examined here, domestic actors initiate international interaction and color its character, but those who present the U.S. position internationally are influenced by concerns at both levels. The way the United States pursues internationalization, however, is almost entirely a product of the interaction among domestic groups. Those who hope to influence international policy would be wise to pay attention to what happens within states as well as between them.

---

[14]  Robert D. Putnam, "Diplomacy and Domestic Politics: The Logic of Two-Level Games," *International Organization* 42 (summer) 1988: 427–60.

# Stage I
# Sources of Internationalization

# 2

# Environmental and Economic Sources of Internationalization

Under what conditions does one state attempt to convince others to adopt regulations it has already undertaken? Which of the environmental regulations the United States imposes on its own citizens does it then also attempt to impose on citizens of other countries? The United States has some of the strictest domestic environmental regulations in the world. Its efforts to make other states adopt similar environmental measures can be seen as ways alternately to improve global environmental responsibility or to impose U.S. hegemony.[1] Regardless, examining which of its domestic standards the U.S. attempts to internationalize can provide important lessons for understanding one of the routes by which international environmental standards are adopted. Also important is that if international regulation comes, at least sometimes, from the domestic regulations of a state that regulates on the issue early, the shape of the international regulation is likely to be influenced by the regulation in the leading state. Knowing which elements of its domestic legislation an environmental leader is likely to work to convince other states to accept can contribute to an understanding of the shape of environmental regulation on the international level.

In chapter 1 *internationalization* as used in this study was defined as a concerted governmental attempt to convince other states to adopt regula-

---

[1] For the more benign view, see, for example, Al Gore, *Earth in the Balance: Ecology and the Human Spirit* (Boston: Houghton Mifflin, 1992), chapter 9; R. Michael M'Gonigle and Mark W. Zacher, *Pollution, Politics, and International Law: Tankers at Sea* (Berkeley and Los Angeles: University of California Press, 1979), pp. 354-55. For a less positive take on U.S. environmental leadership, see, for example, Kazuo Sumi, "The 'Whale War' between Japan and the United States: Problems and Prospects," *Denver Journal of International Law and Policy* 17, no. 2 (1989): 317–72; Peter M. Morrisette, "The Evolution of Policy Responses to Stratospheric Ozone Depletion," *Natural Resources Journal,* 29, no. 3 (1989): 801.

tory standards similar to those the United States already has on a given issue. The focus on governmental efforts ignores phenomena such as voluntary industry standards that may be relevant to international acceptance of environmental protection, but the focus is warranted nonetheless. Governmental policy has behind it the force of law internally and the power of the state externally, and has traditionally been seen as the most important tool for environmental protection. Even if governmental policy does not account for every effort across state lines to persuade others to act environmentally, it is worthy of examination. Moreover, examining governmental activity allows for consideration of the complete set of regulations and resulting internationalization efforts within a given issue area, an analytical process essential for drawing broad conclusions. It is nevertheless useful, even within the cases of governmental action examined here, to be aware of the role that actions by groups that may themselves influence or be influenced by official U.S. action play in support of applying U.S. regulations to other states.

There are two types of policies that are used to pursue the type of internationalization examined in this framework: multilateral diplomacy and unilateral threats. Although different in character, these two types of policies can be similar in result. More importantly, they are often used simultaneously and can thus be seen as two parts of the same strategy. Sanctioning legislation is often considered at the same time as authorization to move forward with multilateral negotiation. If the negotiation fails to produce sufficient internationalization, sanctions can be imposed to require action by other states equivalent to what might have been expected from an international agreement. This two-pronged strategy serves two purposes: it provides an implicit threat ("if you don't go along with an international agreement, we might impose economic harm until you change your policies") and also a backup plan for how to push for internationalization should multilateral efforts prove ineffective. For the purposes of this project, therefore, both unilateral and multilateral actions will be considered subsets of a general strategy to internationalize domestic environmental policy. Further analysis on the different reasons that either strategy prevails might be useful, but it will not be done here.[2]

---

[2] See J. Samuel Barkin and Elizabeth R. DeSombre, "Unilateralism and Multilateralism in Internationl Environmental Politics" (paper presented at the American Political Science Association Annual Meeting, August 1997).

The multilateral path involves a push for international negotiations to deal with a particular issue (usually by creating a new treaty or expanding a current agreement to cover an issue). In the last twenty-five years international environmental diplomacy has been conducted publicly, and generally an attempt to negotiate an international issue has resulted in some sort of agreement. (Judgment is not made at this point about whether such an agreement is effective or largely symbolic.) We can therefore find attempts at internationalization in two ways. One is by examining the existence of treaties on particular issues and looking backward to the negotiation process. The other is by looking at media coverage and governmental reports of environmental negotiations to cover the possibility that negotiations were attempted that did not result in new or expanded treaties. Ascertaining to what extent the United States was the initiator, or simply a participant, in such multilateral efforts could potentially be more difficult. Judgment is made on the basis of U.S. actions and negotiating positions. It is not difficult in practice to tell whether the United States is a leader in efforts at multilateral environmental regulation.

The other path toward internationalization is the use of unilateral threats of restrictions (usually economic) to which the target state will be subject unless it agrees to adopt the policy the United States wishes it to. In practice these threats are of import restrictions, for reasons that will be examined further in chapter 6. It is more difficult to observe threats than attempts at negotiation, because threats can be made in private and need not be implemented. It is nevertheless essential to examine threats, as well as actions taken pursuant to them, because the most effective threats are those that never need to be imposed. At the extreme, threats need not even be made explicit. States may know that certain actions would bring retaliation and therefore they do not even attempt such actions. For example, some states that are not members of the Nuclear Non-Proliferation Treaty may nevertheless refrain from developing nuclear weapons, not because they have been explicitly threatened with retaliation, but because they know that such a threat is implicit. In the cases examined here, however, the United States attempts to gain a specific policy change in the states in question, and that desire must be somehow communicated to the target states. It is therefore possible to observe that communication, particularly since the message is often conveyed through threats of economic sanctions. Import restrictions by the United States are usually authorized by acts of Congress that either impose sanctions or lay out the conditions under which sanctions may or must be imposed. Congress allows the president more latitude in imposing export restrictions while reserving for itself the ability to impose import restrictions, except when the president invokes the cumbersome Emergency Economic Powers Act or the national security provisions of the 1962 Trade Expansion Act.[3] This study therefore looks to congressional legislation for instances of unilateral attempts at internationalization of environmental policy. For the first question posed in this chapter, "Under which conditions does the United States attempt internationalization?" passage of implementing legislation for economic sanctions will serve as evidence of attempts at internationalization.

A second aspect of unilateral threats that must be considered is to what extent this legislation is actually used to attempt to convince states to adopt U.S. standards. International environmental politics is often characterized by symbolic politics, in which states diffuse domestic concern by appearing to take action without actually accomplishing much environmentally.[4] The success or failure of internationalization efforts is considered more fully in chapter 6, but in examining the existence of threats of sanctions we must also look at how serious these threats are. One way to do that is to examine the extent to which the threats that are permitted by domestic legislation are actually made against the target states. Most U.S. environmental sanctioning legislation establishes a process by which sanctions are officially threatened. In some instances an executive branch official is required to certify that a state has not met a required standard. Once a state has been certified, sanctions may be imposed. Certification in these cases functions as an official threat of sanctions, and often suffices to produce a change by the certified states. In other instances, the sanctioning legislation sets out a standard and specifies the states that must

[3] Gary Clyde Hufbauer, Jeffrey J. Schott, and Kimberly Ann Elliott, *Economic Sanctions Reconsidered: History and Current Policy,* 2nd ed. (Washington, D.C.: Institute for International Economics, 1990), p. 66.

[4] Robert O. Keohane, Peter M. Haas, and Marc A. Levy, "The Effectiveness of International Environmental Institutions," in Peter Haas, Robert O. Keohane, and Marc Levy, *Institutions for the Earth: Sources of Effective International Environmental Protection* (Cambridge, Mass.: MIT Press, 1993), p. 18. Bruce Yandle points out that in U.S. congressional policymaking on the environment "symbols are often more important in these struggles than substance." Bruce Yandle, *The Political Limits of Environmental Regulation: Tracking the Unicorn* (New York: Quorum Books, 1989), p. 16.

meet it to export a certain product to the United States. In these cases the legislation itself serves as the threat. Each year, an executive branch office provides a list of the states that have met the standard and may export the product in question to the United States. By these two types of processes, a potential target state is informed at a variety of points that it might be subject to sanctions.

A first way to examine the seriousness of U.S. unilateral attempts at internationalization is therefore to look at both the existence of sanctioning legislation and the instances in which sanctions are officially threatened. The United States has certainly undertaken informal unilateral efforts to convince states to adopt certain environmental policies. These are important and will be discussed, but it is impossible to create an exhaustive list of these efforts. Because sanctioning legislation sets out a process for official threats, it is possible to examine all instances in which threats are made. That would allow for a comparison of areas in which official threats are made or not made, as well as provide a better overall determination of which threats are successful and which are not. Using only official threats is useful because the entire set can be examined. Reliance on official threats, however, will understate the extent of internationalization attempts, since the threat of certification under the legislation can be made unofficially and can also cause policy change. As long as the bias this produces is understood, it should not harm the results of the study.

A second way to assess the seriousness of the U.S. intent of internationalization embodied in these threats is to examine the content of the threat: how serious is the economic harm threatened against the target states? There is no consensus in the literature on economic sanctions about the effects of the sending state's cost of sanctions on their ultimate effectiveness,[5] but it is generally accepted that the more costly sanctions are to the target state, the greater influence they are likely to have.[6] It is important, therefore, to

---

[5] Martin and Baldwin see advantages of costliness of sanctions in demonstrating sending state resolve; Hufbauer et al. point out that expensive sanctions are difficult to sustain and therefore a less credible threat. Lisa L. Martin, *Coercive Cooperation: Explaining Multilateral Economic Sanctions* (Princeton, N. J.: Princeton University Press, 1992); David A. Baldwin, *Economic Statecraft* (Princeton, N. J.: Princeton University Press, 1985). This issue is addressed more fully chapter 6.

[6] See, generally, Hufbauer, Schott, and Elliott, *Economic Sanctions Reconsidered;* Baldwin, *Economic Statecraft.*

examine what economic impact on the target state the sanctions are likely to have. These two paths toward internationalization are considered together. Success of the measures is not considered at this stage except to the extent that the prospect of success bears on the decision to attempt internationalization; success is further examined in chapter 6.

It is also important to note at this point that this book attempts to identify necessary, rather than sufficient, conditions for internationalization of domestic regulations. It examines internationalized regulations to determine what common features they have, without which internationalization would have been unlikely. A complete analysis of every instance in which conditions determined to be necessary might be present in the three issue areas examined, to determine if they will always create internationalization when present, would be unwieldy. The presence of these conditions, however, can be expected generally to lead to internationalization attempts on the part of the sending state.

### Explanations for Internationalization—The Ozone Example

For a variety of reasons states that have passed domestic environmental regulations may want other states to adopt similar measures. States acting alone cannot manage environmental problems that have international causes. They also have little incentive to bear the cost of regulation alone, if the benefits of their environmentalism are not excludable. In some circumstances states may have enough of a net gain from providing public goods that they are willing to regulate unilaterally, but they would nevertheless almost always prefer to have others contribute if possible.

Analyses of the agreements to prevent ozone depletion provide examples of several incentives for internationalization. The explanations examined in this study have been derived in a number of contexts; the use of ozone depletion here is merely as an illustrative example within which to describe the different perspectives, since so many explanations have been given for the origins of this international agreement. First is the "environmental externalities" hypothesis. Because the ozone layer is a true commons and ozone-depleting substances do not affect only the ozone layer over the country that emitted them, it would be impossible for one country alone to protect its share of the ozone layer by unilateral regulations. In fact, any sacrifices made by one state could be negated by the ozone-depleting action of others that did not make similar sacrifices. For environmental reasons alone it was seen as important to gain international regulation of ozone-depleting substances.[7] In the case of ozone layer depletion, several environmental factors that might contribute to internationalization of regulation were present: the environmental damage was potentially severe and decidedly transboundary, and there was (by the time of the negotiation of the Montreal Protocol) consensus on the causes of and solutions to the problem, particularly among those actors in positions to have international influence on the issue.

Another explanation is given, however, for U.S. support of international regulation of ozone-depleting substances. The United States' regulation of some of these chemicals gave the regulated industry a particular set of incentives. A 1977 amendment to the Clean Air Act delegated to the Environmental Protection Agency (EPA) the power to regulate "any substance . . . anticipated to affect the stratosphere" if the effect may "endanger public health or welfare." In 1978 the EPA prohibited the use of chlorofluorocarbons (CFCs) for nonessential aerosols. The Food and Drug Administration (FDA) also passed regulations under the Federal Food, Drug, and Cosmetic Act to prohibit CFCs as propellants in goods, drugs, and medical or cosmetic products manufactured after 15 December 1978. In addition, the Toxic Substances Control Act allowed the EPA to regulate use of CFCs in nonessential aerosols. The FDA and EPA crafted their regulations in a manner so that all CFCs for use in aerosols would fall under one or the other set of regulations.[8]

These regulations gave U.S. industry an incentive to develop substitute chemicals or processes that could take the place of the regulated substances. Doing so, in turn, gave producers of these substitutes an incentive

---

[7] See, for example, Richard Elliot Benedick, *Ozone Diplomacy: New Directions in Safeguarding the Planet* (Cambridge Mass.: Harvard University Press, 1991); Edward A. Parson, "Protecting the Ozone Layer," in Haas, Keohane, and Levy 1993; Elizabeth R. DeSombre and Joanne Kauffman, "Montreal Protocol Multilateral Fund: Partial Success Story," in Robert O. Keohane and Marc Levy, eds., *Institutions for Environmental Aid: Pitfalls and Promise* (Cambridge, Mass.: MIT Press, 1996).

[8] 42 U.S.C. 7457 (b); 21 U.S.C. 301–37; 21 C.F.R. 2.125 (1988); 15 U.S.C. 2601–29; See also Orval E. Nangle, "Stratospheric Ozone: United States Regulation of Chlorofluorocarbons," *Environmental Affairs* 16, no. 4 (1989): 541.

to create an international market for them, through international regulation. The regulations also gave user industries an incentive to make sure that their international competitors were subject to similar regulations. To be prevented from using available, inexpensive, chemicals entailed a cost for the regulated industries that their international competitors were not asked to bear. For reasons of competitive advantage, once U.S. industry was already regulated, industry actors wanted their competitor industries to bear similar regulations. Even Congress noted the competitive advantage issue of domestic regulation and introduced legislation for import restrictions in the event that international regulation was not achieved for ozone-depleting substances. Senator John Chafee (R–R.I.), for example, spoke of the need to "protect our environment and our domestic industry in the near term."[9]

These two explanations certainly intersect, and the United States pursued internationalization of its domestic ozone regulations for a combination of environmental and economic reasons. The impetus for the initial regulation came from environmentalists concerned about the possible hazards to the ozone layer. Richard Benedick, U.S. negotiator for the Montreal Protocol on Substances That Deplete the Ozone Layer, attributes the initial regulation of ozone-depleting substances under the 1977 extension of the Clean Air Act to "public reaction to the revelations of potential danger to the ozone layer."[10] When Congress considered this regulation, industry actors opposed it, arguing that the scientific uncertainty did not justify the economic harm that would befall the regulated user and producer industries.[11] Once regulations had passed, though, repeal looked unlikely.

---

[9] U.S. Congress, *Congressional Record* 8 October 1986, S15679. See also Steven J. Shimberg, "Stratospheric Ozone and Climate Protection: Domestic Legislation and the International Process," *Environmental Law* 21, no. 4 (1991): 2186; and Kenneth A. Oye and James H. Maxwell, "Self-Interest and Environmental Management," *Journal of Theoretical Politics* 6, no. 4, (1995): 607–8.

[10] Benedick, *Ozone Diplomacy*, p. 23; Nangle, "Stratospheric Ozone," pp. 539–40.

[11] See, for example, "DuPont Position Statement on the Chlorofluorocarbon Ozone/Greenhouse Issues," in U.S. Senate, Subcommittees on Environmental Protection and Hazardous Wastes and Toxic Substances of the Committee on Environment and Public Works, *Ozone Depletion, the Greenhouse Effect, and Climate Change*, pt. 2, 100th Cong., 1st sess., 28 January 1987, pp. 170–71, which refers to DuPont's earlier actions against regulation of ozone depleting substances.

Further domestic regulation also looked likely, in the absence of international regulation.[12] At that point, producers and users of ozone-depleting substances joined the call for international regulation, or at least dropped their opposition to it. In 1986, at a critical juncture in the negotiation of the Montreal Protocol, the U.S. industry group Alliance for Responsible CFC Policy publicly supported the idea of international regulation of CFCs.[13] DuPont, the major CFC manufacturer, called for international controls as well.[14] Representatives of these groups and others testified before Congress that unilateral controls by the United States would hurt their business but that they encouraged global action.[15]

To what extent does either explanation, or a combination of the two, explain the internationalization of U.S. domestic environmental policies in general? What is known about the conditions under which either environmental or economic externalities have an influence on the creation of international environmental regulations?

### Environmental Externalities

Much international environmental policy analysis is based on environmental externalities of economic activity. Sulfur emissions from industrial activity in one state drift in the atmosphere and come down as acid rain in another. Pollution that enters rivers in one state flows downstream in the water that runs through another state. When one state's fishers overfish on the open ocean, it affects the future availability of fish for all others.

Not surprisingly, there is a general consensus that human-caused environmental externalities lead to the need to regulate such activities on an inter-

---

[12] Benedick, *Ozone Diplomacy*, p. 31.

[13] Alliance for Responsible CFC Policy, *Press Advisory*, 16 September 1986, as reprinted in Senate Subcommittees on Environmental Protection, *Ozone Depletion, the Greenhouse Effect, and Climate Change*, pt. 2, pp. 176–77.

[14] Senate Subcommittees on Environmental Protection, *Ozone Depletion, the Greenhouse Effect, and Climate Change*, pt. 2, p. 171.

[15] See, for example, testimony of Tover L. Jeansonne, Vice President and General Manager, Diversified Products, Kaiser Chemicals; statement of Harry C. Mandell, Jr., Senior Vice President, Pennalt, Inc.; statement of Richard Barnett, Alliance for Responsible CFC Policy, in U.S. Senate, Subcommittees on Environmental Protection and Hazardous Wastes and Toxic Substances of the Committee on Environment and Public Works, *Stratospheric Ozone Depletion and Chlorofluorocarbons*, 100th Congress, 1st sess., 12, 13, and 14 May 1987, pp. 299–300, 303–4, 319–20.

national level. But how important is this aspect to understanding internationalization of domestic environmental policies? Under what conditions would we expect environmental externalities to play a role in convincing one state to push others to adopt its environmental policies, and what sort of role would we expect them to play?

Some argue that "the case for multilateral action is especially compelling in the case of environmental externalities."[16] Richard Stewart, for example, declares that efforts to harmonize regulations internationally should take place most often for "environmental problems that create especially serious [environmental] externalities." He finds even unilateral trade restrictions justified "to the extent that they are aimed at transboundary pollution or despoliation of the global commons."[17] Certainly many of the analyses of international cooperation on environmental issues, or expansion of U.S. domestic environmental laws, proceed from the assumption that environmental externalities are the driving force for international regulation.

If international environmental externalities are seen as the reason for internationalization of domestic environmental regulations, one can make some further assumptions about the conditions under which one would expect such internationalization to take place. Implicit in Stewart's analysis, for example, are two assumptions: that international action is most needed for environmental problems that are, first, the most serious, and, second, the most transboundary in nature.

Many note the disjuncture in practice between the severity of environmental damage or risk and the propensity of the world to regulate the activity that causes it. Analyses of environmental regulation, nevertheless, often refer to the seriousness of an environmental problem in addressing when it is taken up as an international issue. For instance, Andrew Hurrell and Benedict Kingsbury note that "the perceived seriousness of many environmental problems . . . pushes states toward cooperation and collective management."[18] Detlef Sprinz and Tapani Vaahtoranta hypothesize that "the worse

---

[16]   Richard B. Stewart, "Environmental Regulation and International Competitiveness," *Yale Law Journal* 102 (1993): 2085.

[17]   Ibid., pp. 2098, 2103.

[18]   Andrew Hurrell and Benedict Kingsbury, "Introduction," in *The International Politics of the Environment,* ed. Andrew Hurrell and Benedict Kingsbury (Oxford: Clarendon Press, 1992), p. 14.

the state of the environment, the greater the incentives to reduce the ecological vulnerability of a state."[19] Certainly environmental crises (the discovery of the ozone "hole" or the Chernobyl disaster, for instance) mobilize activity in support of international environmental action. Thus, one would expect internationalization attempts to be strongest where the environmental problems are the worst. Yet the empirical evidence for such a hypothesis alone is weak. Almost any list of the world's worst environmental problems would be different from a list of the environmental issues on which there are international regulatory efforts. Holding constant other variables like the cost of regulation and the states for which the environmental harm is greatest,[20] however, may increase the predictive capability of this formulation.

Another way environmental externalities may impact internationalization comes from the transboundary character of the environmental problem. Transboundary problems are more difficult to regulate than environmental issues that fall within the jurisdiction of only one state because of the difficulties of international cooperation in general.[21] Looking at environmental issues as a whole, one would therefore expect transboundary problems to be addressed less frequently than purely domestic ones. But when states have already regulated domestically, the transboundary nature of environmental problems means that domestic regulations cannot fully address the issue. This inadequacy gives these states a greater environmental incentive to internationalize regulations for environmental issues that have a larger transboundary component. Similarly, Sprinz and Vaahtoranta expect states particularly that are victims, rather than net creators, of transboundary environmental damage (downstream states in river pollution, for example) to push for international controls.[22] If this theory is correct, it would lead us to expect greater effort at internationalization when there is a greater transboundary component to an environmental problem already regulated domestically.

[19] Detlef Sprinz and Tapani Vaahtoranta, "The Interest-Based Explanation of International Environmental Policy," *International Organization* 48 (winter) 1994: 79.

[20] As do Sprinz and Vaahtoranta.

[21] See, for example, Lettie Wenner, "Transboundary Problems in International Law," in *Environmental Politics in the International Arena,* ed. Sheldon Kamieniecki (Albany: State University of New York Press, 1993), pp. 165–66; Matthews, "Introduction and Overview," pp. 32–33.

[22] Sprinz and Vaahtoranta, "Interest-Based Explanation," p. 79.

Attempts to regulate activity relating to environmental externalities, however, have several difficulties. It often is not clear what causes the environmental harm, what its effects are, or how it can be managed. Policymakers in general have a harder time making environmental policy under conditions of uncertainty than when the environmental problem is well-understood. In a strictly functionalist view of uncertainty, the greater the understanding of the causes and effects of and solutions to an environmental problem, the easier it should be to regulate the issue at the international level. This uncertainty factor is most likely to be important when looking at the success of internationalization attempts or at which environmental issues in general have either domestic or international regulations. It is less likely to play a direct role in determining which domestic regulations the United States decides to internationalize, since presumably uncertainty is a less important issue for the state that has already decided to regulate domestically.

The ways in which uncertainty is addressed in environmental politics, however, have an effect on internationalization. One way to address the issue of decision making under conditions of uncertainty is through the idea that "epistemic communities" influence the international adoption of environmental regulations. Epistemic communities are seen as a set of actors with shared knowledge, as well as a set of causal and principled beliefs about an issue; often (although not necessarily) they are comprised of scientists.[23] Peter Haas found, for instance, that in the ozone issue "members of the transnational epistemic community played a primary role in gathering information, disseminating it to governments and CFC manufacturers, and helping them to formulate international, domestic, and industry policies."[24] The actual environmental damage may be less important for this explanation than is the existence of a community of internationally influential actors who can help shape policy on the issue. Haas sees these communities as part of a new pattern of reasoning that governments have learned to apply to environmental policy. He sees international environ-

[23] Peter M. Haas, "Introduction: Epistemic Communities and International Policy Coordination," *International Organization* 46 (winter) (1992): 3.

[24] Peter M. Haas, "Banning Chlorofluorocarbons: Epistemic Community Efforts to Protect Stratospheric Ozone," *International Organization* 46 (winter) (1992): abstract; also pp. 187–224.

mental cooperation as "generated by the influence wielded by specialists with common beliefs," which reflects "a more sophisticated understanding of the complex array of causal interconnections between human environmental and economic activities."[25] This new ecological elite helps national actions addressing environmental problems to converge.[26]

Another similar approach emphasizes information or discourse as the important aspect of knowledge. Karen Litfin, for example, sees scientific culture "as a driving force in the politics of postindustrial society," and examines the power that underlies the use of scientific knowledge.[27] Ronnie Lipschutz sees science and technology as necessary for dealing with problems, but points out that this knowledge is not an absolute and becomes a kind of power or authority that can be used at times when it becomes available or when the political or social structure allows for it.[28] The difficulty with both the epistemic community and the discourse theory responses to uncertainty is that it can be difficult, other than post hoc, to tell whether these actors exist or are influential enough to affect the course of decision making until they do. Such constructivist theories, while perhaps right in essence, are difficult to measure empirically. Nevertheless, we can make some use of the more measurable aspects of these approaches. On the basis of this type of reasoning, one would predict internationalization for environmental regulations when there are internationally connected communities that share an understanding of the environmental issue and have access to and influence with those who would make decisions about internationalization.

In sum, there are a number of instances when, for environmental reasons, a state would want to internationalize its domestic environmental regulations. States may not be able to address serious or transboundary environmental problems without internationalization, and pressure for internationalization may take place through efforts of scientific or environmental groups concerned about these issues.

[25] Peter M. Haas, *Saving the Mediterranean: The Politics of International Environmental Cooperation* (New York: Columbia University Press, 1990), p. xxii.

[26] Haas, *Saving the Mediterranean*, p. 216.

[27] Karen Litfin, *Ozone Discourses: Science and Politics in Global Environmental Cooperation* (New York: Columbia University Press, 1994), pp. 1, 14–51.

[28] Ronnie D. Lipschutz, "Who Knows? Local Knowledge and Global Governance," (paper presented at the 1995 Annual Meeting of the International Studies Association, February 1995, Chicago, Ill.).

Environmental factors are likely to play some role in decisions by the U.S. government to internationalize its domestic regulations. But they are unlikely to play the decisive role, owing both to the problem of social transitivity and the variety of issues that the government has to address at any given time. In the first place, if domestic actors differ in their preference orders for the internationalization of domestic regulations, it is not clear how these differences of preference will be resolved. Arrow's paradox shows that individually transitive preferences will not necessarily lead to clear social preferences.[29] So the simple fact that a number of influential domestic actors are concerned about international environmental protection may not be sufficient to create a domestic push for internationalization. Similarly, even if there are pressing environmental reasons for internationalization of a regulation, there may be intervening governmental priorities other than environmental protection. Thus, while the relative urgency of the problem may influence the willingness of the government, as a whole, to address it on the international level, environmental factors may not be the only cause of internationalization.

### Economic Externalities

Just as there can be environmental impacts from economic activities, there can be economic repercussions from regulations adopted to improve the quality of the environment. The economic externalities, as these repercussions will be called here, result from the differential economic impact that environmental regulation can have on various actors. For example, regulations that impose lower costs on one set of actors than on another set, particularly when the two sets are in competition with each other, are likely to gain the support of the actors who receive lower costs. This set of actors actually can gain comparatively from the regulation. When the regulations affect actors who compete economically on the international level, the analysis of economic externalities can be even more complicated and the results important for the internationalization process. There are two possible economic effects of environmental regulations that could grant competitive advantage incentives for internationalizing domestic regulations: they can result in competitive disadvantages for those

---

[29] Kenneth Arrow, *Social Choice and Individual Values,* 2nd ed. (New York: Wiley, 1963).

who bear the costly regulations, or they can create competitive advantages for those who innovate early in response to domestic regulations when those regulations are internationalized.

We must first examine, however, the relationship between domestic environmental regulations and international competitiveness. There is disagreement about the impacts of domestic environmental regulation. Some see it as harmful to competitiveness, and others see it as beneficial. Costs imposed on a domestic industry but not on its international competitors should, all else being equal, have competitiveness impacts. A first step, therefore, is to ascertain whether or not regulations impose costs overall.

Environmental regulations that increase the costs of operation for an internationally competitive industry reduce its competitive advantage for exports as well as for domestic sales.[30] Environmental regulations can impose a cost on industry either through expenditures for required abatement technology or through higher prices for inputs. Empirical studies have shown that there are costs to industry from domestic environmental regulations, though these are difficult to measure accurately. Gary Yohe demonstrates that "stronger pollution controls do have a backward incidence onto the other factors of production."[31] Anthony Barbera and Virginia McConnell studied the impact of pollution regulation on the growth in productivity of five U.S. industries in the 1970s and concluded that there was a reduction in productivity due to pollution regulation.[32] Joseph Kalt found an inverse correlation for U.S. industries between costs of compliance with domestic environmental regulations and export per-

---

[30] Adam B. Jaffe, Steven R. Peterson, Paul R. Portnoy, and Robert N. Stavins, "Environmental Regulations and International Competitiveness: What Does the Evidence Tell Us?" Draft, 21 December 1993, pp. 5, 14; James Tobey, "The Impact of Domestic Environmental Policies on International Trade," in OECD, *Environmental Policies and Industrial Competitiveness* (Paris: OECD, 1993), p. 48.

[31] Gary W. Yohe, "The Backward Incidence of Pollution Control—Some Comparative Statics in General Equilibrium," *Journal of Environmental Economics and Management* 6 (1979): 197.

[32] Anthony J. Barbera and Virginia D. McConnell, "The Impact of Environmental Regulations on Industry Productivity: Direct and Indirect Effect," *Journal of Environmental Economics and Management* 18 (1990): 62; the decline was due both to the direct costs of pollution abatement technology and to the indirect effects on output.

formance.[33] A 1992 EPA study predicted that, owing to costs imposed by the 1990 Clean Air Act Amendments, the balance of trade in three affected industries (chemical manufacturing, automobile manufacturing, and iron and steel manufacturing) would decline; Dale W. Jorgenson and Peter J. Wilcoxen predicted that GNP would be 3 percent lower in 2005 than it would have been without the amendments.[34] Stewart's analysis of empirical studies on the competitiveness effects of environmental regulations concludes that "environmental regulation has adversely affected U. . productivity."[35]

Regardless of whether environmental regulations may or may not generally disadvantage the economy as a whole, individual industries may be forced to bear higher costs than their competitors owing to specific domestic regulations. It is therefore possible that internationalization could be driven by this competitive disadvantage, with the likelihood increasing the greater the disadvantage (either in terms of severe competitive problems for a small number of firms, or of less severe problems for a larger number of firms).

Others argue that domestic environmental regulation is, at worst, neutral and, at best, beneficial to the economy. Investigations of whether the economic slowdown and trade imbalances in the United States in the 1970s and 1980s could be attributed to the level of environmental regulation show that regulation was not the main culprit.[36] A study of the relationship

[33]  Joseph P. Kalt, "The Impact of Domestic Environmental Regulatory Policies on U.S. International Competitiveness," in *International Competitiveness* ed. A. Michael Spence and Heather A. Hazard (Cambridge Mass.: Ballinger, 1988).

[34]  Carl A. Paskura, Jr., and Deborah Vaughn Nestor, "Environmental Protection Agency, Trade Effects of the 1990 Clean Air Act Amendments," study, 1992; Dale W. Jorgenson and Peter J. Wilcoxen, "Impact of Environmental Legislation on U.S. Economic Growth, Investment, and Capital Costs," in *U.S. Environmental Policy and Economic Growth: How Do We Fare?* ed. Donna L. Bodsky, (Washington, D.C.: American Council for Capital Formation, 1992), pp. 1–39.

[35]  Stewart, "Environmental Regulation," p. 2084.

[36]  Edward P. Denison, *Accounting for Slower Economic Growth: The United States in the 1970s* (Washington, D.C.: Brookings Institution, 1979); J. R. Norsworthy, Michael J. Harper, and Kent Kunze, "The Slowdown in Productivity Growth: Analysis of Some Contributing Factors," *Brookings Papers on Economic Activity* 2 (1979): 387–421; Paul R. Portnoy, "The Macroeconomic Impacts of Federal Environmental Regulation," in *Environmental Regulation and the U.S.*

between the relative stringency of environmental regulations across countries and net exports found no correlation.[37]

To the extent that environmental regulation decreases negative environmental externalities, some argue, it may increase competitiveness in the long term.[38] Increases in water quality standards may, for example, increase productivity in industries that depend on water as an input. Greater health achieved through air pollution regulation may enable workers to be more productive. These results are long-term, diffuse, and difficult to measure, however. More importantly, from a political standpoint, these competitive advantages tend to accrue to diffuse and poorly mobilized actors. They therefore are unlikely to have a direct effect on the likelihood that a state will push for internationalization of a domestic regulation.

There is also some indication that environmental regulations can have advantages to the regulated industry itself. Michael Porter posits that "tough standards trigger innovation and upgrading." On the whole he finds that states with the most stringent requirements often have the lead in exports, even of the regulated product.[39] It could be, however, that the two results are caused by a third variable. It would not be surprising, for instance, to find that states at a high level of economic development have both a lead in exports and a greater degree of environmental protection than those at lower levels.

There could also be a competitive benefit from domestic regulation if environmental rules lead to the use of advanced technologies that are resource

---

*Economy*, ed. Henry M. Peskin, Paul R. Portnoy, and Allan V. Kneese (Baltimore: Johns Hopkins University Press, 1981), pp. 25–54. Gray, however, argues that 30 percent of productivity decline may be attributed to environmental, health, and safety regulations. Wayne B. Gray, "The Cost of Regulation: OSHA, EPA, and the Productivity Slowdown," *American Economic Review* 77 (1987): 998–1006.

[37] James A. Tobey, "The Effects of Domestic Environmental Policies on Patterns of World Trade: An Empirical Test," *Kyklos* 43 (1990): 191–209.

[38] Jaffe et al., "Environmental Regulations," p. 7; Stewart, "Environmental Regulation," p. 2065.

[39] Michael E. Porter, "America's Green Strategy," *Scientific American* (April 1991), p. 168; Michael E. Porter, *The Competitive Advantage of Nations* (London: MacMillan Press, 1990), pp. 647–49.

efficient and therefore result in lower production costs than those absent regulation.[40] Many economists are dubious of this logic, however. If a significant reduction in production cost can be achieved by new technology, we would expect industries to adopt it without regulation; the failure to do so would have to be attributed to noneconomic causes such as bounded rationality or organizational inertia. Another explanation for industry's ignorance of cost-saving improvements might be that a search for efficiency-increasing technology or processes can be costly and therefore is unlikely to be undertaken since it may not produce results. Environmental regulation may provide the incentive to undertake investment that may, in some cases, pay off in an absolute sense. Barbera and McConnell's research shows, however, that the savings due to increased efficiency in production may not always offset the direct costs of the technology.[41]

Another way industry may gain from domestic environmental regulation is in the incentive it gives to industries that make technology used for compliance with the regulation. Particularly in a world in which stricter environmental regulations will be adopted over time, states that regulate their industries early will have a lead in developing and patenting the technology to adjust to the new regulations. If and when other states adopt similar regulations, the industries in the state that regulated first will have a dominant position in "green technology." Theoretically, any business (not simply in the regulated state) could foresee the opportunity to develop environmental technology in the face of regulation, but empirically it does seem that the abatement technology most often appears first in the state that is regulated. For example, the United States leads the world in exports of environmental technology in the areas in which U.S. regulations have been strictest, such as pesticides and hazardous waste management. Germany has the strictest stationary air pollution and water pollution standards, and German industries have a lead in patenting and exporting technology to control air and water pollution.[42]

[40] OECD, *Environmental Policy and Technical Change* (Paris: Organization for Economic Cooperation and Development, 1985).

[41] Barbera and McConnell, "Impact of Environmental Regulations," p. 60.

[42] Porter, "America's Green Strategy," p. 168; Jaffe et al., p. 17; OECD, *The OECD Environment Industry: Situation, Prospects, and Government Policies* (Paris: OECD, 1992).

If there is a first-mover advantage in creating environmental technology, industries in states in which early domestic environmental regulations are passed can benefit. The greater the potential that international regulation provides for sales of domestically developed technology, the more we would expect attempted internationalization of the domestic regulations already in place.

Thus, industries have two incentives to work for internationalization based on economic externalities: to remove the competitive disadvantage they suffer when forced to pay the costs of regulations when their competitors are not, or to gain new markets for regulatory technology to adapt to the regulation. Economic reasons for internationalizing environmental regulations may also not be sufficient by themselves. At any given time most industries would prefer that their competitors either bear costly regulations or be prevented from competing in the same markets in which the domestic industries in question compete. The simple fact that an organized industry group would prefer action be taken against foreign competitors may be insufficient to determine whether such action will be taken.

### Argument: Baptists and Bootleggers for Environmental Protection

In the case of ozone depletion, the environmental and economic externalities explanations intersect through cooperation among the various actors who want internationalization of domestic regulations, albeit for different reasons. The initial domestic regulations on ozone-depleting substances came after pressure on Congress from scientists and environmentalists, and were fought by industry actors, who argued that these regulations were premature and damaging to their operations. Once domestic regulations were in place, environmentalists pushed for internationalization for environmental reasons, and industry eventually pushed for internationalization for economic reasons. Even more interesting is that each group of actors actually made the arguments of the other. In testimony before the Senate Subcommittees on Environmental Protection and Hazardous Wastes and Toxic Substances, which was investigating the possible internationalization of regulations of ozone-depleting substances, the representative of the Natural Resources Defense Council argued that "phase-out would enhance U.S. competitiveness. The incentives created by a reduced supply of conventional CFCs would give U.S. industries the opportunity to steal a march on foreign competitors in the race to commercialize substitutes and recycling techniques."[43] Industry actors were less specific, but did acknowledge the environmental advantages of global action (and the inability of the United States to address the environmental issue through its domestic regulation alone) to preserve the ozone layer.[44] The chair of the industry organization Alliance for Responsible CFC Policy spoke of this coalition of environmentalists and industry actors working together to push for international action on ozone-depleting substances: "It is our belief that the Alliance for Responsible CFC Policy, working in conjunction with the environmental groups, the State Department, and EPA, we can form a very formidable alliance together to impact the world."[45]

This type of coalition has been noted in other issue areas; one analogy is drawn to the coalition of "Baptists and bootleggers" that arose during the time of Prohibition in the United States. At that time, two groups unlikely to be on the same side of an issue found themselves working for the same goal. Baptists favored Prohibition out of a religious and moral opposition to drinking; bootleggers favored it because their profits were higher from selling illegal alcohol to those otherwise unable to obtain it. This analogy, in which an unusual coalition works for the same policy for different reasons, has been applied to a variety of situations, from medieval cloth regulation to the standardization of sizes of loaves of bread.[46]

Bruce Yandle has also used this analogy to characterize policymaking in the United States. He observes, as we see in the ozone debates, that "there

---

[43] Statement of David D. Doniger, Senior Attorney, Natural Resources Defense Council, in Senate Subcommittees on Environmental Protection, *Ozone Depletion, the Greenhouse Effect, and Climate Change,* p. 62.

[44] See, for example, Statement of Robert L. Jeansonne, Vice President and General Manager, Diversified Products, Kaiser Chemicals, in Senate Subcommittees on Environmental Protection, *Stratospheric Ozone Depletion and Chlorofluorocarbons,* p. 299; Statement of Richard Barnett, Chairman, Alliance for Responsible CFC Policy, in Senate Subcommittees on Environmental Protection, *Ozone Depletion, the Greenhouse Effect, and Climate Change,* p. 60.

[45] Statement of Richard Barnett, Chairman, Alliance for Responsible CFC Policy, in Senate Subcommittee on Environmental Protection, *Ozone Depletion, the Greenhouse Effect, and Climate Change,* p. 60.

[46] See, for example, Bruce Yandle, "Intertwined Interests, Rent Seeking, and Regulation," *Social Science Quarterly* (1984): 1002–12.

are ways that one group can use the arguments of the other to achieve totally different objectives."[47] In using the analogy of Baptists and bootleggers, he points out that each of the partners in this type of coalition brings something different to the partnership. In the case of Prohibition, Baptists add "public interest content to what would otherwise be a strictly private venture . . . a moral ring to what might otherwise be viewed as an immoral effort."[48] Bootleggers add economic power and political clout that comes from the concentrated interests they represent.

Kenneth Oye and James Maxwell note that "general environmental concerns are often advanced through the particularistic pursuit of rents of subsidies," and argue that environmental regulations are the most successful when they benefit those actors that are regulated. They argue that when a coalition of self-interested rent-seekers and environmentalists joins together in favor of rent-generating environmental regulation, the coalition is more likely to achieve its goals than either coalition partner acting alone would.[49] Insights from the study of public policy more broadly support these observations. Randall Ripley and Grace Franklin report that industry actors may accept inevitable regulation but "pursue other options designed to make the regulation as light as possible or to acquire governmentally conferred benefits simultaneously as a form of compensation for being regulated";[50] working with those who are the source of the regulation is one option for doing so. Similarly, recent literature in industrial ecology points to many instances in which environmentalists and industry actors can find policies that advance the interests of both groups.[51]

It is reasonable to assume that either group on its own would be less likely to prevail in a domestic political battle than would the two when working

[47] Yandle, *Political Limits of Environmental Regulation*, p. 19.

[48] Ibid., p. 24.

[49] Kenneth A. Oye and James H. Maxwell, "Self-Interest and Environmental Management," *Journal of Theoretical Politics* 6 no.4 (1995): 593, 620.

[50] Randall B. Ripley and Grace A. Franklin, *Congress, the Bureaucracy, and Public Policy*, Rev. ed. (Homewood, Ill.: Dorsey Press, 1980), p. 123.

[51] See, for example, Daniel C. Esty and Michael E. Porter, "Industrial Ecology and Competitiveness: Strategic Implications for the Firm," *Journal of Industrial Ecology* 2, no. 1 (1998): 35–43; R. Socolow, C. Andrews, F. Berkhout, and V. Thomas, *Industrial Ecology and Global Change.* (Cambridge: Cambridge University Press, 1994).

together. In the first place, it is never certain that the U.S. government will attempt to persuade others to adopt its regulations. There is a cost to internationalization. Unilateral measures are often of questionable international legality—for instance, the dispute settlement panel of the General Agreement of Tariffs and Trade held that U.S. sanctions under the Marine Mammal Protection Act on tuna caught in ways that harmed dolphins did not qualify under the GATT's exceptions for environmental protection to its restrictions on unilateral trade measures.[52] Such actions could open the United States to retaliation or to foreign policy failure. Multilateral diplomacy can be costly in terms of the time, effort, and resources devoted to the action. More generally, any state has to be selective in choosing the issues it will put forth as candidates for regulation internationally, particularly when the regulations begin domestically and can therefore be seen as an attempt to impose that state's policies worldwide. Whatever the policy process that operates to produce internationalization, it is therefore unlikely to take up every issue that it is offered.

Environmentalists alone are unlikely to be responsible for U.S. attempts at internationalization of domestic environmental regulation. Empirically, it is clear that not all the issues on the worldwide agenda of any environmental group have been addressed, and it will be shown that the seriousness of the environmental problem or its transboundary nature is not sufficient to lead to internationalization attempts. Nor are industry actors likely to gain their internationalization goals alone. Since their interest is not in worldwide environmental protection but rather in evening the playing field (or tilting it to their advantage), the list of regulations to which they would like to subject their international competitors is likely to be long and not limited to those that benefit the environment most. More importantly, industry actors may prefer protectionism even when it is not intended only as a remedy to competitive disadvantages to which they have been subject; they would sometimes prefer to restrict their competitors in whatev-

[52] General Agreement on Tariffs and Trade, "Protocols, Decisions, Reports 1991–1992 and Forty-eighth Session," *Basic Instruments and Selected Documents,* Supplement No. 39 (Geneva: The Contracting Parties to the General Agreement on Tariffs and Trade, December 1993), pp. 155–205; "United States—Restrictions on Imports of Tuna: Report of the Panel," DS29/R, June 1994. Note, however, that the panel decisions did not have much effect on U.S. policy on tuna issues.

er way possible. Industry calls for economic restrictions on imports from their foreign competitors, couched in environmental protection language, should be seen as another strategy in global economic competition. To the extent that the U.S. government can or will resist general protectionist tendencies, the pleas of industry are more likely to be heard when joined with those of the environmentalists. When such a coalition pushes for an attempt at internationalization, there is little reason not to pay attention to it.

The relationship between environmental and industry actors is not static, however. In the first place, the groups do not exist outside the context in which we examine them. The domestic regulation itself, frequently a result of environmentalist action at an earlier stage, designates and thus constitutes the affected industry. Without an initial regulation there are no affected groups, so to some extent, as is true in the Prohibition analogy, Baptists create bootleggers. It would therefore be a mistake to see them as truly autonomous groups. Second, once the process of internationalization begins it may take on a dynamic of its own. Neither industry actors nor environmentalists may accomplish exactly what they had hoped for. The ozone protection case provides a good example of this phenomenon. Some industry actors pushed for international regulation to help their competitive position and to avoid being subject to even stricter domestic unilateral regulation. Complete phaseout of some classes of chemicals was ultimately adopted in the process that resulted from internationalization efforts, something few industry actors initially foresaw or would have advocated. The environmental actors and the characteristics of the environmental problem itself thus play a central role in the creation of the initial incentives for internationalization and in its changing character.

The existence of such a coalition (different actors pushing for internationalization of domestic regulations for different reasons) can also influence the ways in which internationalization will be attempted. As is examined in more detail in the following chapters, the overlap in the preferences of the coalition members can influence the type of policy the United States pursues. The involvement of industry actors may in fact make unilateral efforts more likely, since threats are likely to be preferred by industry to multilateral diplomacy. Threats of trade restrictions will likely result either in the target state adopting the standards the United States wishes it to adopt, or in restrictions on imports from foreign competitors that have not adopted the standards, the latter of which U.S. industry would in some cases probably prefer.

A question arises as to whether we should take seriously the claims of industries that ask for internationalization of domestic environmental regulations to even the international playing field. Given the previous arguments for competitive advantages from domestic regulations, industry may not in fact suffer from domestic regulations. There are two reasons, however, to consider these industry claims. The first is that the competitive advantages of domestic regulations in several of the above scenarios require that states other than the initial one also adopt similar environmental regulations. The first mover in the development of environmental technology benefits most if there are other states that adopt similar regulatory requirements. New processes that decrease the cost of production are sometimes less costly precisely because widespread regulation has caused the previously required inputs to become more expensive or unavailable. As CFCs were regulated worldwide, for example, they were less available and more costly. Substitutes therefore made production cost less than it would have had CFCs been used, but not less than it would have cost had the world not regulated the availability of CFCs. Adam Jaffe et al., while arguing that overall there is not a radical effect on competitiveness due to environmental regulations, do find that "the degree to which regulatory costs (and benefits) affect trade will depend also on the magnitude of the costs (and benefits) that other countries impose on their domestic industries."[53]

The second reason to consider the concerns of industry actors is that this argument *is* made by the industries in question in the domestic political process that produces internationalization. Stewart points out that in the debate over the competitive advantages or disadvantages of domestic environmental regulations, "people tend to resolve these uncertainties according to their own value orientations and policy predilections." Industry actors who want lower relative regulation and "environmentalists who want to export U.S. standards to other nations, tend to believe that competitiveness effects are large."[54] Whether or not their international competitiveness is indeed harmed, and whether or not they even believe that it is, industry rep-

---

[53] Jaffe et al., "Environmental Regulations," p. 6.
[54] Stewart, "Environmental Regulation," p. 2106.

resentatives use this argument to persuade domestic actors to make policy changes. Generally, the first strategy of industry actors in these cases is to attempt to remove the regulation in question.[55] When that seems unlikely, they argue for its internationalization. When they do so in conjunction with environmentalists, they are likely to convince U.S. policymakers to push other states to adopt these standards.

So, when do we expect the United States to push for internationalization of its domestic environmental policies? Although environmental considerations such as the seriousness of the problem, the transboundary nature of the problem, and the support of the scientific actors are likely to play a role, we should expect a push for internationalization most frequently when domestic regulation causes trade externalities for the regulated industries within the United States. In such situations, a coalition of "Baptists and bootleggers"—environmentalists and industry—forms to persuade the decision makers in question to ensure that other states uphold standards similar to those to which U.S. actors are subject. Alone, either environmentalists or industry actors are likely to be unpersuasive. Together, they often succeed in convincing Congress to pass legislation to work for internationalization of domestic regulations.

The international competitiveness effects of domestic environmental regulations are thus the necessary push, giving the regulated industries incentives to work for internationalization. In cases in which there is a cost to a given industry from the regulation, the industry should want its international competitors to bear the same cost. In the cases in which technology has been produced that responds to the regulation, industry also should want internationalization of the regulation in question to provide further markets for its products. In economic terms these are similar. In both cases there are competitive advantages for domestic industries if industries in other states are subject to similar regulations.

We can then ascertain the conditions under which competitiveness effects lend the incentive for internationalization. First, and perhaps most obvious, is that an industry has to compete internationally for the potential competitive disadvantage of domestic regulations to be an issue. Although not surprising, this observation is nevertheless important because it provides

[55] Stephen M. Meyer, "Environmentalism Doesn't Steal Jobs," *New York Times*, 26 March 1992, p. A23.

a point of comparison with alternate hypotheses about internationalization. Water pollution regulations on purely local lakes, for example, would be unlikely candidates for internationalization if internationalization happens because of environmental externalities. Yet, we would expect pressure for internationalization in that situation, owing to the economic externalities, if the industries that are prevented from polluting the local water compete with industries in other countries that are allowed to pollute their local water supply. Conversely, there might be less pressure than an environmental externalities hypothesis would predict for protecting a species of migratory bird that crosses borders but is not hunted by an organized industry.

A related observation is that environmental regulations on processes are more likely to have harmful competitiveness effects than are regulations on products. When there are domestic regulations on the types of products that can be sold, foreign industries are generally required to meet the same product standards within the state that regulates it, and domestic industries are often not regulated in exports of the products in question. For instance, the U.S. ban on CFCs in aerosols was not specifically directed at U.S. industries, but rather at the aerosols that were sold within the United States.[56] Even if foreign firms meet the same product standards for export to the regulated country, advantages may be gained by the domestic industry in the regulated country (through such things as economies of scale) by having to meet those standards for its domestic market, over the less regulated foreign industry, which has to meet those standards only for goods it wants to export to the state in question.[57] With competition in products therefore less problematic, process regulations are the ones states should try to internationalize. These in turn have implications for the intersection of U.S.-foreign environmental policy and the international trading system.

Finally, it is important that there be an environmental reason to internationalize the regulation. The aforementioned lake pollution example might provide incentives for internationally competing industries to work for analogous regulations for companies in other states not to pollute their lakes, but unless there is a compelling reason that a resource cannot be protected without international action, environmentalists are unlikely to expend

[56] Nangle, "Stratospheric Ozone," p. 540.

[57] Stewart, "Environmental Regulation," pp. 2042–45; see also Porter, *Competitive Advantage of Nations,* p. 47.

their energy working for that goal, and therefore will not lend credence to the internationalization goals of the industry actors. So the importance of international environmental protection of the resource in question does play an important legitimizing role.

## The Cases

The next chapters examine these predictions for U.S. domestic environmental regulations on the issues of endangered species, air pollution, and fisheries regulation. These three issue areas are chosen, in part, because they vary in the extent to which they provide incentives for internationalization for either environmental or economic reasons.

Endangered species regulation seems, at first glance, to provide a number of environmental reasons to internationalize domestic regulations, but few economic reasons. Protection of endangered species has been a centerpiece of the U.S. environmental movement and is an issue about which many feel passionate. At the same time, there is only a marginal domestic industry involved directly in killing endangered species, and even that industry would be harmed more than advantaged by internationalizing the regulation. Internationalized regulations from this area might therefore be expected to demonstrate the importance of environmental factors for determining internationalization.

Air pollution regulations present a more likely avenue for economic-based internationalization. Since many forms of air pollution have mainly local impacts, regulations on such activities would not be candidates for internationalization if environmental explanations predominate. At the same time, domestic air pollution regulations impose serious economic costs on local industries. If economic competitiveness concerns indeed drive internationalization attempts, these regulations would be candidates for such attempts.

International regulation of ocean fisheries is likely to be advantageous for both environmental and economic reasons. As the only true commons among the environmental issues examined here, fishery conservation is simply impossible for one state alone. And economically, fishers bear a high short-run cost from reducing their catch. The extent to which their competitors in other countries do not uphold similar regulations can increase that relative cost.

Comparisons can also be made among the issue areas. If environmental factors matter most, for instance, we might expect those endangered species regulations without direct economic consequences to be subject to the greatest degree of internationalization. Economically driven fishery regulations would be less likely candidates for internationalization, and regulation of air pollution that does not cross borders still less likely ones. Similarly, if economic considerations are the most important, air pollution regulations that harm domestic industries most (or provide the best export opportunities if internationalized) should be much more likely to be internationalized than would regulations protecting endangered species that do not provide immediate economic benefit or loss.

This set of issues is also chosen to be representative of environmental regulations overall. Regulation to protect endangered species often stems from moral concerns about human influences on the natural world. Air pollution regulations address the externalities from industrial activities. Fishery regulations represent resource conservation, the desire to sustainably use a resource so that it can continue to provide goods for human consumption. Among these issues, then, these sets of regulations can be seen to cover a representative range of incentives for environmental regulation.

These three issues are circumscribed enough so that the major U.S. legislation within those issue areas can be examined. Within each issue area, the regulations considered are those that are listed in the *United States Code*, the compendium of U.S. regulations, and the *Code of Federal Regulations*, which contains the details of how those regulations are implemented. The beginning point for these regulations is major environmental legislation passed by Congress, but the implementation of the authority to regulate on a given issue is often carried out by executive branch agencies. It is these agencies that determine, for instance, which species are actually listed under the Endangered Species Act, how the provisions of the Clean Air Act are imposed on polluters, or how international fisheries treaties are implemented on the domestic level.

Evidence of serious efforts to convince target states to adopt the regulations in question is required for a policy to be considered an attempt at internationalization. For that reason, decisions by Congress or the executive orders of the president are the ones that are considered to constitute evidence of attempted internationalization. For each of these regulations an exhaustive examination of congressional legislation and executive orders has been undertaken, as well as a search of the *Federal Register*, which lists all policy actions taken by U.S. agencies in response to domestic regulations, so that the list of internationalization attempts discussed here represents a complete list of measures the United States has attempted to internationalize. Even international negotiation almost always involves creation by Congress of the authority for U.S. negotiators, so it is likely that examining congressional and executive documents leads to a nearly exhaustive list of efforts to push environmental regulation internationally. It is possible that U.S. policy makers may informally suggest to states that they adopt regulations. Although this study notes those processes, it does not claim to have gathered a complete list of such attempts. If anything, then, this study underestimates the degree of attempted internationalization. Of greatest importance, however, are those efforts that the United States is willing to undertake on a sustained level, and for that to happen a legislative decision must be taken; this study examines the full set of these internationalization attempts.

U.S. action on these issues shows that internationalization of domestic regulations is most frequently attempted when the adoption of U.S. standards by other states will increase the competitive advantage of U.S. industries, and when both industry and environmental actors work for internationalization. Although the importance of environmental and economic incentives for internationalization vary across cases, the set of regulations the United States decides to push on the international scene are those for which there are both economic and environmental concerns.

# 3

# Endangered Species

Regulations to protect endangered species whose ranges or migratory patterns cross national borders are logical candidates for international attention. National protection of such species will have little effect if members of the species are unsafe when they cross a border, or if the species range primarily outside the state that regulates their use. Conservation of endangered species was one of the early environmental issues for which there were attempts at international cooperation. There were informal bilateral agreements between the United States and Russia to regulate sealing as early as 1893, and the multilateral North Pacific Sealing Convention was negotiated in 1911. Early agreements to protect migratory birds include a 1936 agreement between the United States and Mexico. Other early agreements include the 1946 International Convention for the Regulation of Whaling (which had its origins in earlier whaling agreements) and the 1940 Convention on Nature Protection and Wildlife Preservation in the Western Hemisphere. Nevertheless, regulation of endangered species often begins (and sometimes ends) within one state.

The United States has an active history of domestic attempts to preserve wildlife threatened with extinction. U.S. legislation to protect threatened species began in 1900 with the Lacey Act.[1] This act prohibits trafficking in animals killed illegally in their state of origin, but also prohibits importing, exporting, transporting, or selling wildlife taken in violation of any U.S. law, treaty, or regulation.[2] Currently, the main U.S. legislation relating to endangered species is the Endangered Species Act (ESA).[3] Many of the regulations

---

[1] Ch. 553, section 1–5; 31 Stat. 187.

[2] 16 U.S.C. 3372 (a) (1) (1988).

[3] P.L. 93–205; 87 Stat. 884 (1973), plus later amendments.

protecting threatened species are tied directly to this act, either as amendments or as separate legislation justified by the requirements under the act to protect species. Considered here as candidates for internationalization are the Endangered Species Act (including its amendments and precedent acts) and the Marine Mammal Protection Act (MMPA),[4] each of which is implemented through additional legislation as well. Marine mammals listed under the Endangered Species Act are automatically subject to the terms of the Marine Mammal Protection Act, though marine mammals that are not listed under the ESA are also given protection under the MMPA. The species regulated under these two acts are examined here to determine which ones the United States pushes to protect on the international level. To limit the complexity of the survey, only animal species are considered. It is also important to examine the details of which species are protected and in what way. To differentiate among potential explanations for internationalization we need to know not only that aspects of the Endangered Species Act and related regulations have been internationalized, but *which* aspects. Does the United States push harder for foreign protection of certain species? Are certain ways of regulating species protection more likely to be internationalized?

**Endangered Species Act**

The history of the Endangered Species Act began with the Endangered Species Preservation Act of 1966,[5] which regulated native U.S. fish and wildlife threatened with extinction and began the process of governmental listing of species at risk. The Endangered Species Conservation Act of 1969 expanded species protection to include wildlife not native to the United States and strengthened the provisions for listing endangered species.[6] The current (amended) version of the general species protection legislation is the Endangered Species Act of 1973.[7] This legislation applies to all endangered or threatened plants and animals. It regulates not only taking or importation of listed species, but also exports and sales, as well

---

[4] 86 Stat. 1027, 16 U.S.C. 1361–1407, P. L. 92–522.

[5] P.L. 89–669; 80 Stat. 926 (1966).

[6] P.L. 91–135; 83 Stat. 275 (1969); it was an amendment to the 1966 Act.

[7] P.L. 93-205; 87 Stat. 884 (1973).

as other activities. Amendments in 1978, 1979, 1982, and 1988 have preserved the basic framework of the act but refined the details and have generally extended the ability of the government to regulate activities that harm threatened or endangered species. Current debates about the authorization of the ESA could change the provisions of the legislation dramatically.

The main provisions of the act relate to the "listing" of species that are at risk. Depending on the species in question, the secretary of the interior or of commerce has jurisdiction for determining whether a species is endangered or threatened. The secretaries work, respectively, through the U.S. Fish and Wildlife Service or the National Marine Fisheries Service. (The secretary of agriculture is involved for the listing of plant species.) The types of eligible species have increased with each change in legislation. Initially the act regulated only native fish and wildlife. The expansion in 1969 included worldwide species, and the one in 1973 included plants and invertebrates. Under the 1973 legislation (the most recent legal definition), a species is considered "endangered" if it is in danger of extinction through all or a significant part of its range.[8] It is considered "threatened" if it is likely to become endangered within the foreseeable future.[9] Wildlife can also be listed as endangered or threatened within a specific geographic area. There are different implications of the two definitions. The act forbids government actors or private parties to harm or traffic in endangered (but not in threatened) species. It forbids the federal government to take action that would harm either endangered species or degrade their critical habitat. The legislation authorizes but does not require the secretaries to extend this latter provision to cover threatened species as well. The secretaries are also permitted to list species as though they were endangered or threatened, even if they are not, if they are sufficiently like those that are imperiled that enforcement officials would have difficulty telling the difference between the imperiled and nonimperiled species.

Consideration of a species for listing can begin either at the initiative of the responsible secretary or by petition from an interested person.[10] The secretary also refers to listings under the Convention on International Trade in Endangered Species of Wild Fauna and Flora (CITES) or to listings by

[8]  16 U.S.C. 1532 (6) (1988).
[9]  16 U.S.C. 1532 (20) (1988).
[10]  16 U.S.C. 1533 (b) (3) (A).

state agencies within the United States as evidence that species are imperiled, although neither automatically leads to listing on the federal level. The secretary in question then examines the species to see if it meets the criteria for listing, set out in section 4 of the act. If the secretary decides that a species should be listed, notice of such is put in the Federal Register and given to interested groups and individuals, and the secretary will hold a hearing on the subject if requested to do so. There are provisions for emergency listings, but these listings are in effect for only a limited time unless the standard listing procedure has been completed within that time. The status of separate breeding populations of species that have been listed is required to be reviewed every five years. The ESA also requires that a recovery plan be prepared for each listed species. Listing is time consuming and there is a serious backlog of species under consideration at any time. More than one thousand species are currently listed as endangered or threatened.

The Endangered Species Act as a whole has been one of the most controversial pieces of legislation in U.S. history. Initially, however, the act passed without much resistance, because it seemed to be largely symbolic and unlikely to involve high costs.[11] It soon became obvious that the costs would indeed be high for some sectors of domestic industry. The history of the act has involved near-constant efforts to repeal it, and some efforts, as well, to apply its provisions internationally.

### Marine Mammal Protection Act

Congress passed the Marine Mammal Protection Act in 1972 "to prohibit the harassing, catching, and killing of marine mammals."[12] As the domestic environmental movement in the United States grew, whales, dolphins, and other large marine mammals became rallying cries and symbols of the need to restrain indiscriminate environmental destruction.

The act pertains to the protection of all marine mammals, including whales, seals, polar bears, and sea otters. In general it requires that the taking of any marine mammal comes only by permit after an appropriate authority has been convinced that the taking will not disadvantage the

[11] P. Korn, "The Case for Preservation," *Nation*, 30 March 1992, pp. 415–17.

[12] Legislative History, p. 4144, from House Report (Merchant Marine and Fisheries Committee) No. 92–707, 4 December 1971.

species. Most attention in the act is given to the protection of dolphins. Among other things, the act aims to reduce "the rate of incidental kill or serious injury of marine mammals . . . to insignificant levels approaching a zero mortality and serious injury rate."[13]

To this end, the National Marine Fisheries Service is authorized to require fishing vessels to take various measures to limit the numbers of marine mammals killed or injured incidental to commercial fishing operations, and to regulate the ways in which marine mammals are captured or harassed for other reasons. These regulations have changed over time, but have included such things as requiring observers on board fishing vessels, legislating the use of certain types of fishing and safety gear, and limiting the numbers of marine mammals that could be taken for display in zoos or aquariums. After 20 October 1974 no marine mammals could be taken "in the course of a commercial fishing operation unless the taking constitutes an incidental catch" and is covered by a permit to allow incidental taking.[14] Permits were issued either to specific commercial fishing operations or to a class or organization of commercial fishers.

Unlike the ESA, the MMPA was clearly understood at passage by those obligated to uphold it to be costly. The final version of the legislation was significantly weakened from the original proposal to ban the taking of marine mammals altogether, but even the weaker provisions were likely to require difficult changes for some who relied on the hunting of marine mammals, either directly or indirectly. The MMPA, too, was subject to efforts to eliminate or weaken some of its provisions, as well as to apply them internationally.

### Internationalization of Endangered Species Regulations

The United States has attempted to internationalize endangered species regulations both generally and with respect to specific species. Domestic actors have lobbied Congress to pursue international regulation of the issues relating to general endangered species protection, as well as worked specifically for the international protection of elephants, sea turtles, dolphins, and

---

[13] P.L. 92–522; Section 101 (a) (1) (B).

[14] 39 *Federal Register* 32118–9. The *Federal Register* will henceforth be cited as [vol.] *FR* [pages].

whales. Species protection in general can be a high-profile environmental interest. The species on whose behalf the United States worked hardest to increase international protection are certainly ones about which there are high levels of public concern.

Depending on the species and the form of regulation in place, some efforts at expanding protection of endangered species internationally required negotiation of international agreements; for other internationalization efforts Congress passed new legislation or modified the regulations being internationalized. In still other instances Congress modified internationalization efforts created in a different context. The legislation used for this latter effort was the Pelly Amendment to the Fisherman's Protective Act, passed in the context of fisheries legislation and discussed further in chapter 5.

## CITES

In the general issues of species protection, the United States was one of the main actors behind the negotiations to create the Convention on International Trade in Endangered Species of Wild Fauna and Flora (CITES). The Endangered Species Conservation Act of 1969 directed the government to negotiate an international treaty to protect endangered species. Section 5 of the act calls for the United States to "encourage foreign countries to provide protection to species and subspecies of fish or wildlife threatened with worldwide extinction" and to seek "the signing of a binding international convention on the conservation of endangered species."[15] The actual conference that produced the convention took place in Washington, D.C., in 1973 and was hailed by environmental organizations as crucial to the protection of species internationally. Once the treaty entered into force, attention shifted to convincing states with large markets for endangered species, such as China and Japan, to join. Environmental organizations such as the World Wildlife Fund and the then International Union for the Conservation of Nature and Natural Resources joined in lobbying trips to these states to help persuade them of the benefit of joining the agreement.[16]

[15] See P.L. 91–135, Section 5 (a) and (b).
[16] "Senior World Wildlife Fund Officials..." *Reuters,* 2 October 1979 (Lexis/Nexis); Graham Earnshaw, "China Today..." *Reuters,* 23 September 1979 (Lexis/Nexis).

Congress also passed several pieces of general legislation to allow prohibition of trade in endangered species and to allow other types of import restrictions in cases in which states are found to be "undermining the effectiveness" of international endangered species programs. The 1969 ESA itself began the prohibition of importation of species considered endangered.[17]

Conservation organizations suggested the concept of expanding existing fisheries internationalization legislation to include other aspects of wildlife protection. The precedent had been set to hold states to existing international standards for the protection of fisheries, and the modification of existing legislation would be easier than the creation of new instruments for internationalization. The Defenders of Wildlife, the Society for Animal Protective Legislation, the Humane Society of the United States, the International Fund for Animal Welfare, the Fund for Animals, the International Primate Protection League, and the Washington Humane Society all testified before Congress in favor of a tool that would allow for trade sanctions against states that did not protect endangered species.[18] These groups pointed out that although the United States was a signatory to CITES, it had no ability to regulate illegal wildlife trade except that coming into the United States. Such a measure, its supporters argued, would strengthen CITES.[19] Legislation to restrict wildlife imports from states that did not protect endangered species had few opponents. Such a measure was likely to be applied in mostly symbolic ways, and it would certainly not hurt U.S. domestic interests to convince others to protect endangered species more thoroughly. Moreover, the regulations proposed were discretionary.

Congress passed such an amendment to the Pelly Amendment in 1978. It leaves the determination of whether a state does not adequately protect endangered species up to the secretary of commerce or the secretary of the interior. Upon finding "that nationals of a foreign country, directly or indirectly, are engaging in trade or taking which diminishes the effectiveness of any international program for endangered or threatened species," the relevant secretary certifies that the state is doing so. This certification, in

---

[17] P.L. 91–135, Section 2.

[18] Legislative History, p. 1773; from House Report No. 95–1029, pp. 10–11.

[19] Ibid., pp. 1774–75.

turn, gives the president discretion to cut off all wildlife imports from that state to the United States.[20] When it seemed likely that simply cutting off wildlife imports from an offending state might not be sufficient protection, the same set of actors worked for a 1992 amendment that expanded the sanctioning provisions so that certification for endangered species issues could also incur import restrictions on fish, as the other provisions of the Pelly Amendment could.[21]

In addition to focusing this legislation to protect specific endangered species internationally, the United States applied these provisions to some states that did not uphold CITES export provisions in general. At the insistence of domestic environmental organizations, their first use was in the protection of sea turtles. The Interior and Commerce departments certified Japan for diminishing the effectiveness of international treaties protecting endangered species, because of its trade in endangered sea turtles.

A more elaborate use of this legislation was made in an effort to protect tigers and rhinoceroses, which are protected under both CITES and the ESA. There is no inherent connection between these species, except that they are often used in traditional medicine in the same countries. The United States cited the People's Republic of China and Taiwan several times under the Pelly Amendment wildlife provisions for diminishing the effectiveness of CITES by trading in tiger and rhinoceros parts. The United States also targeted Singapore for its unrestricted trade in rhinoceros parts.

In addition to general efforts to protect endangered species internationally, the United States passed separate legislation to work specifically for the internationalization of regulations to protect elephants, sea turtles, and dolphins. It also applied general legislation to specific protection of whales.

### Elephants

Under U.S. law—the Endangered Species Act and the domestic application of CITES regulations—killing of elephants is regulated. The ESA lists African elephants as threatened.[22] Prior to 1989, trade in ivory in the United States was permitted as long as that trade was conducted under CITES regulations.

---

[20] P.L. 95–376.

[21] 106 Stat. 4903; 22 U.S.C. 1978.

[22] 50 CFR 17.11.

In 1977 CITES listed the African elephant on Appendix II, which allows for regulated trade in endangered species and their parts. CITES also designed particular control measures for ivory from African elephants in 1985. Beginning in 1989, the organization moved the African elephant to Appendix I, thereby calling for an end to commercial trade in elephant parts, including ivory. At the 1997 CITES Meeting of the Parties, the African elephant stocks in Zimbabwe, Botswana, and Namibia were downlisted to Appendix II, allowing for controlled trade beginning in 1999.[23] The Asian elephant is listed on Appendix I of CITES and was listed there before its African counterpart; it is also listed as endangered under the ESA. Female Asian elephants do not have visible tusks and the species is therefore less subject to—though not immune from—poaching for ivory. Since there are no wild elephants in the United States, the only domestic regulations that have any real impact domestically are those that prohibit trade in ivory.

Protection of elephants has long been a concern of environmental organizations in the United States. Elephants are considered to be noble creatures worthy of admiration. They are the largest of all land animals and are considered to live in families.[24] Environmental organizations have used them on posters and in fundraising appeals.

Because the African elephant is listed as threatened under the ESA, the secretary of the interior is given the power to issue regulations for its protection. These regulations included a prohibition on imports of ivory from states that were not parties to CITES, and the requirement that legal imports be accompanied by export permits that indicate the country of origin of the ivory and any intermediary states through which it has passed; all of these states must be members of CITES.[25]

Domestic environmental organizations lobbied Congress in the 1970s and 1980s to achieve greater protection of elephants internationally. A bill to prohibit ivory imports from states with inadequate elephant protection was introduced into Congress in 1979 by Representative Anthony C. Bielenson

[23] D'vora Ben Shaul, "High Stakes," *Jerusalem Post,* 21 September 1997, p.7 (Lexis/Nexis).

[24] World Wildlife Fund, "A Program to Save the African Elephant," *World Wildlife Fund Letter,* no. 2, 1989, pp. 1–2.

[25] *Endangered Species Act Amendments,* P.L. 100–478; "Endangered Species Act Amendments," House Report No. 100–827, p. 2721.

(D-Cal.) but failed in the Senate. Several proposals were introduced into Congress in the 1987–88 session. One proposal was to prohibit all exports of ivory from states not party to CITES and from those states that do not have native populations of African elephants. Another included a proposal for a tax on imported ivory. The most restrictive proposal would have prohibited the United States from importing or exporting any elephant products. Representatives from environmental organizations such as the World Wildlife Fund and the Humane Society of the United States testified in favor of added restrictions on the ivory trade, though with differences of opinion as to the likely effects of the various options.[26]

The eventual result of these lobbying efforts was the 1988 African Elephant Conservation Act, actually a part of the Endangered Species Act Amendments of 1988. This legislation establishes "a moratorium on ivory trade with countries not having effective elephant protection programs." The U.S. government was required to review the African elephant conservation programs for all ivory-producing countries within a year after the effective date of the legislation and to determine whether those countries met certain criteria. A state that wants to trade in ivory must be a party to CITES and adhere to the CITES Ivory Control System. It must have an elephant conservation program based on the best available information and must make progress in compiling relevant information; it must effectively control and monitor the taking of elephants. In addition it must not allow the export of ivory in excess of its ivory quota under CITES.[27] If a state does not meet all these criteria, the "Secretary shall establish a moratorium on the importation of raw and worked ivory" from that country. If there was not sufficient information to make a determination by 31 December 1989, the secretary was to impose a moratorium on importing ivory beginning 1 January 1990, until enough information was collected to conclude that the state meets the criteria.[28]

The Pelly provisions to work for internationalization of wildlife protection discussed earlier were also amended in 1988 to address the adherence

[26] Ibid., p. 2725–27. Some discussed fears that stricter restrictions on imports from intermediary states without bringing primary ivory exporters into agreements regulating ivory would make elephant conservation more difficult.

[27] P.L. 100–478 Sec. 2201 (b) (1); 102 Stat. 2318.

[28] P.L. 100–478 Sec. 2202 (a) (1); Sec. 2202 (2); 102 Stat. 2319.

of countries to CITES regulations on elephant conservation. If a country does not adhere to the CITES Ivory Control System, "that country is deemed for purposes of section 1978(a)(2) of title 22, to be diminishing the effectiveness of an international program for endangered or threatened species" and subject to the same types of import restrictions allowed under the Pelly Amendment in general.[29]

These pieces of legislation are still in effect, but as long as CITES enforces a ban on ivory trading they have little practical use. Now that the new CITES elephant regime for Southern African countries has taken effect, this legislation may once again be relevant.

### Sea Turtles

All species of sea turtles are listed as endangered or threatened under the Endangered Species Act. Leatherback and hawksbill turtles were listed as endangered in all of their range on 2 June 1970. The Kemp's Ridley turtle was listed as endangered 2 December 1970. The green turtle was listed as threatened on 28 July 1978 (except for breeding populations of Florida and the Pacific Coast of Mexico, which were listed as endangered). The loggerhead turtle was listed as threatened (throughout its range) on 28 July 1978.[30] In October 1987 the Department of Commerce, as authorized by the Endangered Species Act, passed regulations requiring that U.S. shrimp trawlers in the Gulf of Mexico and the Atlantic Ocean off the southeastern United States reduce sea turtle mortality from shrimp trawling. To reduce the effects of shrimping on sea turtles, shrimp trawlers were originally required to use "turtle excluder devices" (TEDs) on their shrimp trawl nets between 1 May and 31 August, beginning in 1989. These TEDs are like trap doors that allow turtles, which would drown when caught in shrimp nets, to escape. Trawlers under twenty-five feet in length could meet the requirements instead by limiting their tow times to ninety minutes. The dates between which the regulations were in effect varied by region.[31] The law was expanded in 1992 to require use of TEDs year-round. This

[29] P.L. 100–478, Title II, Sec. 2303, Oct. 7, 1988, 102 Stat. 2322. 16 U.S.C. Sec. 4242.

[30] National Research Council, *Decline of the Sea Turtles: Causes and Prevention* (Washington, D.C.: National Academy Press, 1990), p. 16.

[31] *52 FR 24244–62.*

legislation also removed the exemption for trawlers that limited their tow time to 90 minutes.[32]

Reaction by U.S. shrimp fishers to sea turtle protection measures was swift and negative. From the beginning sea turtle protection was targeted mainly at industry actors rather than at habitat protection or other conservation measures. Shrimpers argued that the devices themselves, which cost approximately $600 each, cut into their profits and caused them to lose up to 50 percent of the shrimp they caught.[33] In the summer of 1989 an organization calling itself the Concerned Shrimpers of America organized a blockade of shipping lanes along the gulf coast by shrimp vessels in protest of the new requirements, and threatened violence against enforcement officials. Shrimpers picketed George Bush in September of that year during his visit to the area. Others undertook acts of civil disobedience, or simply refused to adopt the new technology.[34]

In response to protests, Secretary of Commerce Robert Mossbacher suspended the implementation of the domestic TED requirement in July 1989, allowing alternate fishing practices that might protect sea turtles. The National Wildlife Federation, an environmental nongovernmental organization (NGO), challenged the suspension in federal court, which ordered the Department of Commerce to implement the TED requirement. It did so in September.[35]

Industry actors put pressure on Congress to change the law. They argued that the use of TEDs was hurting their livelihood, particularly in comparison with foreign shrimpers who did not have to use the devices. When the

[32] 57 FR 18446ff.

[33] Most studies put the number closer to 10 percent, and later studies showed that TEDs did not preceptibly decrease shrimp harvest. Jack Rudloe and Anne Rudloe, "Shrimpers and Lawmakers Collide Over a Move to Save the Sea Turtle," *Smithsonian*, 20 (December) 1989, pp. 44ff; "U.S. Tells Shrimpers to Give Sea Turtles an Escape 'Door,'" *New York Times*, 6 September 1989, p. B8 (Lexis/Nexis).

[34] Anthony V. Margavio and Craig J. Forsyth, *Caught in the Net: The Conflict between Shrimpers and Conservationists* (College Station, Tex.: Texas A&M University Press, 1996), pp. 31–34.

[35] Eugene H. Buck, "Turtle Excluder Devices: Sea Turtles and/or Shrimp?" *CRS Report for Congress*, 28 November 1990, p. 9; Patty Curtain, "Annual Shrimping Kill: 44,000 Turtles," *St. Petersburg Times*, 18 May 1990, p. 1A (Lexis/Nexis); Jeff Klinkenberg, "Stubborn Shrimpers May Face Consumer Backlash over Turtles," *St. Petersburg Times*, 8 April 1990, p. 5D (Lexis/Nexis).

regulations were adopted, foreign shrimp accounted for approximately 73 percent of U.S. shrimp consumption.[36] The president of the Texas Shrimp Association testified before Congress that foreign shrimpers, without having to bear the cost of buying and using TEDs, were "dumping" shrimp in the U.S. market, "affecting the price the American producer receives."[37] In congressional debate on the issue, representatives from shrimping areas indicated that "it would be an outrage if this country imported shrimp from countries like Mexico who do not utilize these turtle-excluder devices while our shrimpers are being penalized." Others argued that "if we must use TEDs then everybody else ought to have to use TEDs as well. . . . [I]f they are not required to do that which we are required, then we should not be required to import their shrimp."[38]

At the same time, environmentalists were also pushing for broader adoption of the TED requirement. The Earth Island Institute organized an advertising campaign pointing out that "Mexico's shrimp fleet is killing an estimated 11,000 sea turtles on the Endangered Species Act."[39] The chair of the wildlife program of the Environmental Defense Fund testified before Congress that "we are very sensitive to the concerns of the shrimpers with respect to foreign competition,"[40] and argued in favor of holding all shrimping states to the same standards required of U.S. shrimpers.

Responding to industry and environmentalist pressure, Congress passed Section 609 of Public Law 101-162, taking effect November 1991. This legislation directs the Department of State to "initiate negotiations . . . for the protection and conservation of [United States–regulated] species of sea turtles." It also called for economic measures to be taken against states that

---

[36] Buck, "Turtle Excluder Devices," p. 2.

[37] Statement of Harrus Lasseigne, Jr., U.S. Congress, House, Subcommittee on Fisheries and Wildlife Conservation and the Environment of the Committee on Merchant Marine and Fisheries, "Sea Turtle Conservation and the Shrimp Industry," 101st Congress, 2nd Session, Serial No. 101–83, 1 May 1990, p. 13.

[38] U.S. Congress, *Congressional Record,* Senate, 101st Congress, 1st Sess., 29 September 1989, p. s12266, Statements of Mr. Johnson and Mr. Breux.

[39] Earth Island Institute, "Can We Stop Mexico's Shocking Slaughter of Endangered Sea Turtles?" Mailing, no date, also in newspapers as ads.

[40] Statement of Michael Bean, U.S. Congress, House, Subcommittee on Fisheries and Wildlife Conservation and the Environment of the Committee on Merchant Marine and Fisheries, "Sea Turtle Conservation and the Shrimp Industry," p. 18.

did not apply the same level of sea turtle protection that the United States mandated. The secretary of state was directed to provide Congress with a list of all states conducting shrimp fishing within the range of endangered sea turtles in ways that could harm them. The law calls for a prohibition on the imports of shrimp or shrimp products "which have been harvested with commercial fishing technology which may affect adversely such species of sea turtles," no later than 1 May 1991. This import ban does not apply, however, in cases where the president certifies to Congress (annually) that the government of the state in question has adopted a regulatory program comparable to the U.S. regulations governing incidental taking of sea turtles; that the average rate of incidental take of sea turtles by vessels of the state is comparable to the U.S. average incidental take; or that "the particular fishing environment of the harvesting nation does not pose a threat of the incidental taking of such sea turtles in the course of such harvesting."[41]

Domestic battles ensued over the implementation of the internationalization legislation. The Department of State initially decided to apply the standards only to a group of fourteen states that catch shrimp in the Caribbean and Western Atlantic. Environmental NGOs, led by Earth Island Institute, sued the Department of Commerce to expand the regulations to apply to additional states, as well. In December 1995, the U.S. Court of International Commerce ruled that the regulations must be applied to all states that catch shrimp, thus expanding the number of states subject to the certification process to seventy.[42]

The United States also took part in the negotiation of the Inter-American Convention for the Protection and Conservation of Sea Turtles (the Salvador Convention), agreed to in 1994 and as of this writing open for signature.[43]

## Dolphins

Under the Marine Mammal Protection Act, U.S. citizens are not allowed to take or possess any marine mammal for which they have not been granted

[41] P.L. 101–162, §609, (b) (2).

[42] *Earth Island Institute et al. v Warren Christopher et al.*, United States Court of International Trade, Court No. 94–06–00321, 913 F. Supp. 599; Humberto Marquez, "Shrimp, Next on the List for U.S. Decertification," *Inter Press Service*, 25 April 1996 (Lexis/Nexis).

[43] "Countries Agree on World's First Sea Turtle Treaty," *Marine Conservation News*, 8, no. 4 (1996): 1, 16.

a permit, or "to transport, purchase, sell, or offer to purchase or sell any marine mammal or marine mammal product" or "use in a commercial fishery any means or method of fishing in contravention of regulations and limitations issued by the Secretary of Commerce for that fishery" to protect marine mammals.[44] Although the only dolphins currently listed in the Endangered Species Act are those found in China and Pakistan, the National Marine Fisheries Service determined that the northern offshore spotted dolphin and eastern spinner dolphin were "depleted."[45]

Most dolphins are killed in a process called "setting on dolphins." For reasons that are still unclear, yellowfin tuna in the Eastern Tropical Pacific Ocean (ETP) school with dolphins. Tuna fishers can spot dolphins as they surface to breathe, encircle them with large purse-seine nets, and collect the tuna below. In the process, dolphins are trapped under water and drown. The National Marine Fisheries Service (under the auspices of the Department of Commerce) began regulation of dolphin mortality in 1974 by setting an annual quota of seventy-eight thousand dolphins that could be killed in the process of tuna fishing. This quota was to be lowered annually. The regulations also included restrictions on gear and the requirement that fishing vessels encircling dolphins use a difficult and time-consuming method of "backing down" to allow dolphins to escape.[46]

In the early 1970s approximately 75 percent of the tuna catch, by more than seven thousand tuna fishers in the United States, came from the use of purse-seine nets. In that same period annual dolphin deaths in the ETP numbered above three hundred thousand.[47] U.S. tuna fishers were predictably concerned about a regulation that would require them to change their fishing practices. Tuna fishers claimed the industry "could not survive economically" under MMPA restrictions.[48] The president and past presi-

[44] 50 C.F.R. 18.13.

[45] *57 FR* 47620–24.

[46] National Research Council, *Dolphins and the Tuna Industry* (Washington, D.C.: National Academy Press, 1992), p. 27.

[47] The United States was responsible for at least 85 percent of these deaths. Alessandro Bonanno and Douglas Constance, *Caught in the Net: The Global Tuna Industry, Environmentalism, and the State* (Lawrence: University Press of Kansas, 1996), pp. 121, 125, 127.

[48] "Porpoise-kill Limit Set by House," *Facts on File World News Digest,* 11 June 1977, p. 442D3 (Lexis/Nexis).

dent of the American Tunaboat Association as well as other leading U.S. tuna fishers said that the MMPA provisions would put them out of business and threatened to register their boats in foreign countries to avoid the costs of the regulation.[49] The Pacific tuna fleet staged a strike for three months in 1977 as a protest against the regulations. Tuna fishers "are madder than hell," noted Stan Levitz, President of the Master Mates Association.[50] Many noted the increasing number of dolphin deaths caused by countries whose tuna competes with that from the United States. As a story in *Forbes* put it, "the crowning irony of the tuna drama is the same one the U.S. faces in many other cases in which its environmental concerns are not shared by other countries: Foreign fishermen are not subject to U.S. law; they're operating as always."[51]

Environmentalists were concerned as well. They argued for stronger domestic regulations that would lower the number of dolphins permitted to be killed each year, and they went so far as to sue the National Marine Fisheries Service for failing to protect dolphins sufficiently as required by the MMPA. They also pointed out that increasing numbers of dolphins were killed in foreign tuna-fishing processes, and reflagging by U.S. fishing vessels would only increase the risk of dolphin deaths worldwide.

In response to these concerns, Congress included provisions in the MMPA from early on that were also directed toward other states. In the first set of marine mammal regulations promulgated by the National Marine Fisheries Service under the MMPA it became illegal to import any fish into the United States if "such fish were caught in a manner prohibited by these regulations or in a manner that would not be allowed" by a person under the jurisdiction of the United States.[52] All imported fish, except those determined by the director of the NMFS to be uninvolved with commercial fishing operations that injure or kill marine mammals, thus had to meet certain requirements. The director of the NMFS could find that fishing operations of the state in question are "accomplished in a manner which

---

[49] "American Tunaboat Association . . ." *New York Times*, 1 January 1977, p. 6 (Lexis/Nexis).

[50] Dan Tedrick, "American Tuna Fleet . . ." *Associated Press*, 3 May 1977 (Lexis/Nexis).

[51] Bruce Coleman, "Troubled Waters," *Forbes*, 1 April 1977, p. 56 (Lexis/Nexis).

[52] 39 *FR* 32124.

does not result in an incidental rate in excess of that which results from fishing operations under these regulations," even if they are not in conformity with the specific U.S. regulations. After this finding is published, it must be accompanied by a statement from a "responsible official" of the country of origin attesting either that the fish were caught in a manner consistent with the regulations or that they were caught in a way that did not result in the killing of or injury to numbers of marine mammals in excess of those permitted under the United States regulations.[53] In December 1975 the director of NMFS determined that the fish subject to regulations would consist of yellowfin tuna, salmon, halibut, and pilchards (from South Africa). These fish were prohibited from entering the United States unless the above conditions had been met. The regulation of salmon, halibut, and pilchards could be met by a statement by either the master of the vessel that caught the fish or a responsible official of the country of origin that those particular fish were not caught "in a manner prohibited for U.S. fishermen." Also acceptable was certification by the country in question that "all of the vessels under its flag are fishing in conformance with these regulations." Although the legislation provided additional hoops to jump through for a state that wished to export fish to the United States, states fishing for salmon, halibut, and pilchard generally did not have to change their fishing behavior, just document that they were not harming marine mammals.[54] The most important target of the import restrictions under the Marine Mammal Protection Act has been yellowfin tuna.

The MMPA requires that all yellowfin tuna imported into the United States be accompanied by a Yellowfin Tuna Certificate of Origin from the country whose flag vessels caught the tuna, including a statement that the tuna was caught "in conformance with the United States Marine Mammal Regulations."[55] In addition, imports of yellowfin tuna and tuna products from states known to fish in the Eastern Tropical Pacific Ocean (whether the particular tuna was caught there or elsewhere) are contingent upon a favorable finding by the assistant administrator for fisheries of the NMFS. The assistant administrator requires that the fishing practices of the state in question "are conducted in conformance with U.S. regulations and stan-

---

[53]  39 *FR* 32124.

[54]  40 *FR* 56904; 42 *FR* 64559.

[55]  42 *FR* 64559.

dards" or that "such fishing is accomplished in a manner which does not result in an incidental mortality and serious injury [of marine mammals] in excess of that which results from U.S. fishing operations under these regulations."[56] To receive a finding from the assistant administrator of fisheries, the countries of origin must submit a variety of information, about such things as the fishing technology used and the country's procedures to protect marine mammals, as well as copies of laws and regulations that will protect marine mammals.[57]

Congress has amended the MMPA several times. In response to pressure from domestic fishers concerned about competitive advantage and to appease environmentalists who did not believe sufficient dolphin protection was being applied, some of these amendments related to import restrictions on tuna caught by foreign fishers. The 1984 amendments required that states exporting tuna to the United States document that they had adopted dolphin conservation programs that were equivalent to those followed by the United States, and with comparable average dolphin mortality rates. The 1988 amendments set the "comparable" rate at 1.25 times that of U.S. dolphin mortality, and required in addition that any intermediary exporter of tuna to the United States certify and provide proof to the United States that it has prohibited importation of tuna from states from which the United States has already embargoed tuna.[58] Congress also amended the Pelly Amendment to the Fisherman's Protective Act so that states subject to import restrictions under the MMPA for six months would automatically receive Pelly certification, which would allow the president to prohibit all imports of fish or fish products from those states.[59]

The implementation of many of these regulations was fraught with domestic battles as well. Although Congress passed the law and its amendments, the National Marine Fisheries Service of the National Oceanic and Atmospheric Administration (under the Commerce Department) passes and applies the regulations to implement it. Congress passed the 1984 and 1988 amendments, in part, because of dissatisfaction with the extent to which NMFS was applying the restrictions to foreign fishers. Earth

[56] 42 FR 64558.

[57] 42 FR 64559–60.

[58] P.L. 100–711.

[59] 22 U.S.C. 1978 (b).

Island Institute, a U.S. environmental organization, sued the Department of Commerce to force it to restrict imports of tuna under its own regulations. The original embargo of Mexican tuna and the expansion of the embargo to other tuna-fishing states were imposed by court order. Similarly, import restrictions against tuna from "intermediary" states—those that buy tuna from embargoed states—were ordered and then expanded after the Earth Island Institute and other environmental organizations won their lawsuit forcing the United States to expand the list of states considered intermediary tuna exporters. Later, the definition of *intermediary* was changed (with the approval of Earth Island Institute), which removed most intermediary states from the list.[60]

Domestic consumers were also engaged in their own nongovernmental form of internationalization, as they began to demand "dolphin-safe" tuna. Some tuna canneries, sensing a market niche, already provided labels indicating that the tuna they marketed had been caught in ways that did not harm dolphins.[61] Environmental organizations lobbied Congress, which ultimately passed the Dolphin Protection Consumer Information Act in 1990, to create standards tuna would have to meet in order to be considered "dolphin-safe." This action happened alongside official governmental action in favor of internationalization of MMPA regulations and was certainly related to it. Some of the same environmental groups were working both to influence consumer demand domestically and to impact governmental decisions to require foreign dolphin protection policies. In addition, U.S. tuna fishers would gain competitively by such labeling and by consumer demand, since their tuna was already "dolphin-safe" and would therefore enjoy a marketing advantage.

Congress also directed the United States to take a variety of multilateral actions relating to internationalization of the MMPA. The legislation directed the United States to begin "negotiations immediately to encourage the development of international arrangements for research on, and conservation of, all marine mammals."[62] The legislation provided a specific plan for the secretaries of interior and state to follow, beginning with negotiations

---

[60] *Earth Island Institute et al.* v *Mosbacher et al.,* United States District Court for the Northern District of California, No. C 88 1380 TEH, 785 F. Supp. 826, 3 February 1992.

[61] National Research Council, *Dolphins and the Tuna Industry,* p. 32.

[62] 16 U.S.C. 1379 (b) (1) (1988).

with foreign governments having jurisdiction over commercial fishing fleets whose operations harm marine mammals.[63] The secretaries were also instructed to encourage treaties that would address tuna-fishing effects on marine mammals and that would protect ocean and land regions important to marine mammal populations, as well as to work for the amendment of existing treaties to make them compatible with the act. The legislation in addition instructed them to initiate a multilateral conference for the purpose of obtaining an international convention on marine mammal conservation and protection.[64] The United States also opened negotiations to organize a dolphin protection regime within the Inter-American Tropical Tuna Commission, and between the IATTC and other states. Several international agreements have been successfully concluded and have entered into force, as discussed further in chapter 7.

**Whales**

The Marine Mammal Protection Act also outlawed all commercial whaling for U.S. whalers. Moreover, most species of whales are listed as endangered under the Endangered Species Act. And because the United States is a member of the International Whaling Commission (IWC), regulations passed by that organization also become regulations under U.S. domestic law. Under IWC regulations, commercial whaling was legal but subject to restriction on the size (and species) of whales caught, the seasons and areas in which they could be caught, and the overall numbers of whales harvested. In 1986 the organization implemented a prohibition on commercial whaling agreed to in 1982.

After the passage of the MMPA the United States sought almost immediately to internationalize its prohibition against commercial whaling to the extent possible. In the interim it also sought to apply IWC whaling restrictions to all whaling states. Because the U.S. whaling industry, already almost nonexistent by the time of the MMPA, was effectively ended by its passage, there was little economic reason to push whale conservation internationally. But there were clear environmental reasons. The world's whaling stocks were dwindling, and whales had always been a symbol of the environmental movement, more even than elephants.

[63]  16 U.S.C. 1379 (c) (1988).
[64]  102 Stat. 4765.

In the case of whaling, internationalization was not difficult. Environmental organizations used existing legislation, applying an aspect of the Pelly Amendment referred to in the discussion of general endangered species internationalization strategies and discussed further in chapter 5. The legislation refers to states that "diminish the effectiveness of an international fishery conservation program," and allows the president to restrict fish imports from these states. In the legislation the term "fish products" is defined to mean "fish and marine mammals," and "international fishery conservation program" is defined as "any ban, restriction, regulation, or other measure in force pursuant to a multilateral agreement to which the United States is a signatory party, the purpose of which is to conserve or protect the living resources of the sea."[65] Because of this broad definition, the United States was able to work for internationalization of U.S. regulations against whaling under legislation already passed for other purposes.

Additionally, because the discretionary nature of the Pelly Amendment meant that the president was not required to impose sanctions on states certified for whaling, Congress in 1979 passed the Packwood-Magnuson Amendment to the Magnuson Fishery Conservation and Management Act (FCMA) of 1976. The FCMA requires, among other things, that states that wish access to the United States Fishery Conservation Zone must sign fishery agreements with the United States and be given specific fishery allocations. This system set the stage for the United States to deny access to its fisheries to gain policy changes in the countries that wanted to fish in U.S. waters.[66] Under Packwood-Magnuson, certification by the Department of Commerce under the Pelly Amendment that a state was "diminishing the effectiveness of the IWC" triggered certification under the Packwood-Magnuson Amendment as well. Unlike the voluntary nature of those under the Pelly Amendment, however, sanctions under Packwood-Magnuson are automatic. Once a state is certified, the secretary of state is required to

---

[65] P.L. 92–219, sec. 8, (g) (3) and (4).

[66] The first denials of access to fishing rights in U.S. waters were imposed in ad hoc ways to influence policy that had nothing to do with conservation policy. The United States reduced and then eliminated the allowed Soviet catch in response to the Soviet invasion of Afghanistan in 1979. The United States suspended Poland's fishing rights following the 1981 repression of the Solidarity movement. David D. Caron, "International Sanctions, Ocean Management, and the Law of the Sea: A Study of Denial of Access to Fisheries," *Ecology Law Quarterly* 16 (1989): 315.

reduce that state's fishing allocation in U.S. waters by at least 50 percent. If by a year later the conditions that led to certification are not corrected, no fishing rights will be granted to the state in question. With this legislation in support of existing methods of internationalization, the United States was able to push whale conservation internationally.

## Endangered Species Conclusions

Under pressure from domestic actors, the U.S. government thus decided to push endangered species regulations internationally. Official, as well as unofficial, government actions suggested both general regulations to limit trade in endangered species (with a particular focus on rhinoceroses and tigers), and specific international regulations to protect elephants, sea turtles, dolphins, and whales. It is immediately obvious that none of the environmental or economic explanations alone account for the pattern of attempted internationalization observed. At first glance even the Baptist and bootlegger theory seems suspect. What explains the pattern of attempted internationalization that emerges?

### Environmental Externalities

Environmental concerns were the driving force behind the original endangered species regulations, but environmental factors alone do not explain the internationalization attempts the United States makes in the area of species protection. If the U.S. government attempts to persuade others to adopt United States–style regulations for environmental reasons alone, we would expect the United States to work for the internationalization of rules to protect species that are the most endangered, the most important, or whose range is the most transboundary. A comparison of these environmental externalities explanations for internationalization with the regulations the U.S. government pushes internationally indicates that environmental exigencies cannot be the only explanation for internationalization efforts.

If internationalization were pursued for the species that are the most endangered, regulations pertaining to species listed as endangered under the ESA should be better candidates for internationalization than those that pertain to species listed simply as threatened. We might also expect regulations pertaining to species with the lowest population numbers to be the best candidates for internationalization. Some of the species the U.S. works hardest to protect internationally, such as rhinos, tigers, and elephants, are seriously threatened with extinction. Some of the most endangered species, however, are ignored in the process of internationalization, for example, the giant panda, the Asiatic black bear, and the Saigo antelope. At the same time, some of the species the United States works to protect internationally are not seriously threatened. Dolphins, for example, one of the species for which the United States has worked hardest for international protection, are not even listed as endangered under the ESA or CITES, and were only marginally threatened when the domestic regulations began. Some whale species are or have been seriously threatened, but some, like the minke currently taken "scientifically" by Japan and commercially by Norway, are now abundant.

If internationalization of regulations is pursued for species that cannot be protected without specifically *international* regulation, we should expect the United States to push for international protection of species that have explicitly transboundary ranges. Species found in the ocean, especially those that travel long distances, are clear candidates for internationalization on the basis of transboundary issues. Birds that travel from one area to another would also be important. Other candidates for internationalization owing to environmental necessity would be regulations pertaining to species that exist primarily outside U.S. jurisdiction and therefore not subject to protection by legislation aimed at U.S. actors.

The transboundary nature of the species does have a high correlation with attempted internationalization, since several of the efforts at internationalization involve species with particularly great transboundary ranges, such as sea turtles, dolphins, and whales. Other species with transboundary ranges, such as birds, are not protected, however. Some of the species protected, such as elephants, rhinos, and tigers, do not frequently travel across borders, but they are resident entirely outside the United States and therefore in need of international rather than domestic protection.

The need to protect some species may be more critical, because of the environmental effects that their extinction would cause, than the need to protect other, less important, species. The theory of "keystone" species posits, for instance, that some species are particularly important to their ecosystems.

We might expect internationalization of regulations of those species that are seen to be more important for their ecosystems, regardless of how threatened those species are individually. A number of the identified keystone species are plants, not considered in this study. But some animal species have also been identified as essential to their ecosystems. African elephants are seen as important to maintaining the balance among plants and animals in their ecosystem,[67] and they were the subject of regulatory internationalization efforts. Beavers, central to the maintenance of wetlands, and gophers, which play the same role for meadows,[68] were not the subject of such efforts. Large carnivores are often particularly important for maintaining balance in the food chain of their ecosystems,[69] and many of the species protected through internationalization efforts fit that category, though perhaps more for reasons discussed below.

Another way to look at the environmental importance of species is to examine those species that environmental organizations work hardest to preserve. These species may not be the most endangered, but environmentalists may nevertheless work hardest for their preservation because these species will be the ones for which international regulations are most likely, or because efforts to preserve them may be important in gaining support for the general idea of preserving endangered species. These species could thus be seen as "keystone species" in the effort to preserve endangered species in general. Some refer to them as "flagship species" and note that they are almost always "charismatic megafauna";[70] species that look good on World Wildlife Fund calendars. Several influential wildlife organizations in the United States and elsewhere publish lists of the most endangered species. The World Wildlife Fund's 1994 Ten Most Endangered list, for instance, includes the following animal species: tiger, black rhinoceros, giant panda, Asiatic black bear, hawksbill sea turtle, Saigo antelope, Egyptian tortoise,

---

[67] See Michael J. Glennon, "Has International Law Failed the Elephant?" *American Journal of International Law* 84 (1990): 7; C. Paine, "A Note on Trophic Complexity and Community Stability," *American Naturalist* 103 (1969): 91.

[68] Reed F. Noss, "From Endangered Species to Biodiversity," in *Balancing on the Brink of Extinction: The Endangered Species Act and Lessons for the Future,* ed. Kathryn A. Kohn (Washington, D.C.: Island Press, 1991), pp. 233–34.

[69] Ibid., p. 235.

[70] Ibid., p. 235.

and the red and blue lory.[71] The World Conservation Monitoring Center's top-twenty endangered species list explicitly acknowledges the aim "not to devise a definitive 'Global Top Twenty,' but a selection of those animals in danger of extinction which are of popular interest." This list includes the following species: numbat, golden bamboo lemur, woolly spider monkey, kouprey, Mediterranean monk seal, Yangtze River dolphin, black rhinoceros, Javan rhinoceros, giant panda, scimitar-horned oryx, kakapo, Seychelles magpie robin, Chinese alligator, and Kemp's Ridley sea turtle.[72] Ocean dolphins and whales have also been the focus of concern on the part of nongovernmental organizations.

Some of the species that scientists or environmental organizations generally worked to protect ended up on the list of those subject to internationalized regulations, but not all. The one obvious trait these species share is that they are large and photogenic; all of the individual species the U.S. pushed to regulate internationally are species that environmental organizations use as "flagship" species. But that is clearly not a sufficient explanation, since many of the species on the "top ten" or "twenty" lists are not subjects of official U.S. pressure for international protection, despite being threatened with extinction.

Not surprisingly, all the species the United States decided to work to internationalize regulations for were a priority for one or more environmental reasons, but looking at these reasons alone would not allow us to derive a list of attempted regulatory internationalization. There must be factors other than environmental necessity that contribute to the likelihood that regulations for species will be internationalized.

### Economic Externalities

Likewise, economic factors alone do not explain the pattern of internationalization observed. If they did, we would expect the United States to

---

[71] World Wildlife Fund, "WWF's 1994 Ten Most Endangered List," http://envirolink.org/arrs/endangered.html, date visited: 9 March 1996. In the United States this organization is still callled by its original name, "World Wildlife Fund"; internationally it is now known as "The World Wide Fund for Nature."

[72] World Conservation Monitoring Center, "Global Top Twenty," http://wcmc.org.uk/infoserve/species/sp_top20.html, date visited: 11 June 1996. The species given as examples here are those that do not exist only within the United States.

work to internationalize those domestic environmental regulations that have competitive advantage effects on U.S. domestic industries. The economic effect on industries caused by endangered species regulations (either negatively, by those who lose from the domestic legislation, or positively, by those who develop and want to sell technology to conform with the regulation) would predict the extent to which such legislation will be internationalized.

There is certainly some effect of international competition on the species regulations the United States tries to push internationally. If internationalization attempts come in cases where there are positive impacts on U.S. actors when others protect endangered species, we would see the United States pushing for international regulations when U.S. industries gain from an increased market for technology they produce to meet the obligations. At first glance it might not seem that endangered species regulation is likely to have a technological aspect that can create a competitive advantage for U.S. businesses, but when we consider the regulations on economic activities that harm endangered species as a side effect, it becomes apparent that there are some technological aspects to complying with endangered species regulations.

Advantage to those who manufacture technology, however, is not a good predictor of internationalization of species regulations, since few involve technological solutions and the one in this case that does (protection of sea turtles through the requirement of turtle excluder devices on shrimp trawl nets) is so small and inexpensive that it is unlikely to provide a sufficient incentive for manufacturers to press for regulations. The required use of turtle excluder devices domestically for shrimp fishing, for instance, is surely a boon for manufacturers of these devices, (almost all of them in the United States)[73] who gain if their devices are more widely required. The more than five-hundred-dollar price for a TED may be prohibitively high for some shrimp fishers forced to purchase one or several. But the small number required overall and the small size of the U.S. TED industry makes competitive technological advantage an unlikely explanation for internationalization, even of sea turtle protection. This type of explanation, therefore, does not contribute much to an understanding of the

---

[73] William Schomberg, "Mexico: Prawns in the Export Game," *El Financiero International,* 17 May 1993, p. 10 (Lexis/Nexis).

internationalization of endangered species regulations. It may, however, contribute to the understanding of internationalization in other, more technologically based environmental problems.

If the United States works to gain international adherence to regulations that are costly to U.S. actors competing internationally, we would expect the extent to which U.S. industries bear costs from endangered species regulations relative to their foreign competitors to determine the extent to which these regulations are internationalized. At first glance this explanation for U.S. international efforts on behalf of endangered species seems unlikely as well. By the second half of the twentieth century any U.S. industry that benefited directly from harming endangered species was minuscule, and there is unlikely to be international competition in the killing of endangered animals. U.S. wildlife trade is not insignificant, however. In the 1980s U.S. imports and exports of wildlife products were worth more than $962 million annually.[74]

We should expect little economic impetus for hunters, however, to push for international regulation of the species they are prevented from hunting in the United States. If anything, these actors would probably want regulation to remain at domestic levels to preserve the possibility that they could continue their activity in other countries. Yet, even though there are few industries in the United States that depend directly on the exploitation of endangered species themselves, protection of species can nevertheless have enormous impact on industries. For example, the battle between the timber industry in the northwestern United States and ESA protection of the spotted owl reached the Supreme Court. Earlier, concern about the fate of the endangered snail darter prevented the completion of an enormous Tennessee Valley Authority dam project in Tennessee. Economic activity that harms endangered species as an externality of its main purpose is vulnerable to endangered species regulation. Note that these two controversial species protection issues involved the necessity of protecting the habitat of endangered species. If reasons of competitive disadvantage are to be the

[74] U.S. Fish and Wildlife Service, *International Trade in Animal Products Threatens Wildlife* (Washington, D.C.: U.S. Fish and Wildlife Service, 1981), cited in Simon Lyster, *International Wildlife Law* (Cambridge: Grotius Publications, 1985), p. 239. See also Sarah Fitzgerald, *International Wildlife Trade: Whose Business Is It?* (Washington, D.C.: World Wildlife Fund, 1989), p. 9.

source of pressure for internationalization of species regulation, habitat protection regulations are likely to be the ones pushed internationally. Although the project interrupted by the snail darter is not in an industry competing internationally, forestry products (the problem in the spotted owl's habitat) are internationally traded commodities. So if the states internationalize regulations because of competitive disadvantage, we might expect to see pressure for other states to adopt habitat protection for endangered species.

Also important for internationalization for reasons of competitive disadvantage would be those species regulations that are targeted at industries that kill threatened species in the process of their intended activity. Tuna fishers forced to find ways to avoid killing dolphins lose the most cost-effective and efficient method of fishing. Shrimpers must buy new equipment for their boats and catch shrimp in a way that may allow some to escape. Tuna- and shrimp-fishing industries are huge international endeavors. Both of these regulations would consequently be subject to industry pressure for internationalization based on competitive disadvantage.

Competitive disadvantage for U.S. industry actors resulting from endangered species regulations clearly provides some impetus for internationalization of these rules. Not only were regulations pertaining to dolphins internationalized when dolphins are not seriously endangered, but the only aspects of the MMPA dolphin regulations the United States worked to internationalize were those relating to the incidental catch of dolphins in tuna-fishing activities, not those addressing harassment or capture of dolphins for aquariums. Some elements of U.S. actions, however, remain unexplained by this approach. Since habitat protection, particularly relating to forests and wetlands, is among the most costly of endangered species regulation, we would expect these types of regulations to be pushed internationally. In addition, protection of some species, like elephants, rhinos, and tigers, causes little if any harm to domestic economic interests. So although the economic externalities of environmental regulations provide some impetus to convince other states to adopt similar regulations, they do not tell the whole story.

### Baptists and Bootleggers

The pattern of attempted internationalization observed in addressing protection of endangered species regulations becomes clear when we consider the intersection between economic and environmental goals. Both are present in the jockeying over endangered species regulations. Environmental interests begin the process of domestic regulation, even more in this issue area than in others. The protection of endangered species rarely has a direct economic benefit, so industry actors are not particularly involved when regulations are proposed. They become engaged afterward, along lines determined by how the initial regulations are created. Measures to protect dolphins and sea turtles could have been created in a number of different ways. It was because they were first aimed at the domestic fishing industry that actors in that industry became involved in the process.

It was the involvement of these industry actors that created the necessary push for internationalization. The regulations of most concern to industry actors—those protecting dolphins and sea turtles—also concerned environmentalists who wanted to broaden the protection for these species. U.S. fishers (of tuna and shrimp), as well as representatives of environmental groups, testified before Congress in favor of internationalizing dolphin and sea turtle protection. A Baptist-bootlegger coalition is apparent even in the issue of consumer demand for "dolphin-safe" tuna as a form of internationalization. Environmental groups advocated green consumerism as a way to protect a resource, and U.S. tuna fishers saw it as a way to increase the demand for the version of a good in which they had the competitive advantage.

More interesting, though, is that the conjunction of Baptist and bootlegger interests can be seen even in areas where there would not appear to be clear economic interest. The section of the 1969 Endangered Species Act that called for international negotiation of a treaty to regulate species protection, for instance, did so "to assure the worldwide conservation of endangered species and to prevent competitive harm to affected United States industries."[75] Likewise, the instrument most frequently used to attempt to internationalize U.S. species regulations (specifically those relating to whales, elephants, rhinoceroses, tigers, and species protection in general, and expanded to apply to dolphin protection as well) is the Pelly Amendment. The Pelly Amendment was passed with a Baptist-and bootlegger coalition of fishery organizations and environmentalists described further in chapter 5.

[75] P.L. 91–135, Section 5 (a) and (b).

Essentially, then, although environmentalists were able to broaden existing legislation to work for the internationalization of endangered species more generally, they had success in doing so only when building on existing legislation that was passed by a coalition of environmentalists and industry actors. In the effort to internationalize whale conservation the current legislation could be used (and propped up through supporting legislation negotiated later). In other instances environmentalists were able to influence the modification of legislation for broader use than originally intended. But even when their concern was not related to industry, they managed to implement regulations addressing their concern only with legislation that was created with industry interest.

It is also telling to examine which elements of endangered species regulations the United States decided to push internationally. All the regulations examined had elements relating to habitat protection that in the long run might provide more important protection of species than some of the measures that were internationalized. With respect to sea turtles, it is likely that more harm comes to the species from lack of adequate protection of nesting sites or from trade in animal parts for traditional medicine than from shrimp fishing. But although some of the trade aspects have been addressed through CITES-related internationalization, international protection of sea turtles was pursued primarily through restrictions on foreign fishers. Not surprisingly, it is these types of restrictions from which the U.S. shrimp-fishing industry would benefit.

The attempt to expand domestic legislation to address international endangered species regulation cannot be sufficiently explained by characteristics of the environmental problem alone, nor by industry concern over competitive advantage. It is the cases in which the concerns of the two groups align in favor of internationalization that the United States is willing to attempt to persuade other states to adopt analogous regulations. And if the presence of industry actors is important to push forward international regulations on an issue as lacking in industry interests as endangered species regulation can be, their presence in a coalition with environmentalists is likely to be even more important in the internationalization of other environmental regulations.

# 4

# Air Pollution

Regulation of air quality is a useful issue area through which to examine the extent and success of internationalization. Some aspects of air pollution are inherently transboundary in character, since air currents do not stop at state borders. Not all air pollution, however, has transboundary environmental effects. Much of the pollution emitted into the air stays over, or is deposited on, the country from which it originates. One of the reasons to examine air pollution, then, is that it does not necessarily have transboundary environmental effects.

Air pollution regulation is, moreover, an issue that interests both environmentalists and industry actors. The reasons for their involvement are clear. On the one hand, air quality has long been recognized as a health and environmental issue; it has been regulated in the United States locally for more than a century, and nationally since the 1960s.[1] On the other hand, the regulations that have been passed to address air problems have had clear and generally negative effects on polluting industries.

Almost all national air pollution regulations in the United States have their origins in the Clean Air Act Amendments of 1970 and 1990. Between these two versions of the act, air pollution regulations can be seen to consist of six basic types: motor vehicle regulations (including restrictions on types of fuel sold as well as specifications on the manufacture of automobiles), control of the release of toxic air pollution, reduction in sulfur dioxide emission to prevent acid rain, preservation of the ozone layer,

---

[1]  Reiner Lock and Dennis P. Hartwick, eds., *The New Clean Air Act: Compliance and Opportunity* (Arlington, Va.: Public Utilities Reports, Inc., 1991), pp. 3–4.

ambient air quality regulations, and performance standards for new pollution sources.[2]

## Clean Air Act

Although there have been local air pollution regulations in the United States for more than a century, national air pollution policy began in 1955 with the "Air Pollution, Control—Research and Technical Assistance" law.[3] That act provided more coordination than regulation, however, and it was not until the Clean Air Act of 1963 that air pollution was regulated at the national level.[4] Amendments in 1965 enabled the government to set automobile emissions standards for carbon monoxide and hydrocarbons.[5] The 1967 amendments authorized the regulation of ambient air quality.[6] The legislation took on its current form beginning with the 1970 Clean Air Act Amendments.[7]

Richard Nixon proposed updating the 1967 act at the beginning of 1970, as part of a broader push for new environmental regulations. Most of the industries facing new or stricter regulations knew little could be done to avoid them altogether and instead worked to steer the process toward regulatory outcomes they would prefer. Auto industries, for instance, fought with the majority actors in the House and Senate over whether emissions standards would be written into law or set by the EPA, with the belief that the latter would be more susceptible to political pressure after the law was enacted.

The newly amended act required the government to set National Ambient Air Quality Standards (NAAQS) that states would be required to achieve, and that would be used to limit individual source emissions. Congress gave the newly created Environmental Protection Agency responsibility for oversight, and for stepping in to implement the act in any U.S. state that

---

[2] See "Summary of S. 1630," *Congressional Digest* 69, no. 3 (1990): 71–72, 96. Not all provisions are in all versions of the act.

[3] P.L. 84–159; 69 Stat. 322 (1955); see also Lock and Hartwick, *New Clean Air Act,* p. 4.

[4] P.L. 88–206; 77 Stat. 392 (1963).

[5] P.L. 89–206; 79 Stat. 992 (1965).

[6] P.L. 90–148; 81 Stat. 485 (1967).

[7] P.L. 91–604; 84 Stat. 1676 (1970). This is the initial version of the act that most people refer to as the Clean Air Act.

did not adequately do so.[8] The 1970 act also created technology standards for new pollution sources and regulated the release of toxic air pollution.[9] And, importantly, it expanded the process of regulating emissions from mobile sources, such as automobiles.[10]

As the 1977 deadline for a number of air quality standards approached, it became clear that states within the United States would not be able to meet the standards set by the act in 1970.[11] Congress scrambled to produce regulatory proposals that would continue to move forward the goal of clean air while revising the requirements. The fact that standards would not be met opened the door to discussion of what other processes might be mandated to improve air quality. Foremost among these was the issue of requiring flue gas desulfurizers, or scrubbers, to clean $SO_2$ from emissions at factories and power plants. It was in the context of this policy consideration that a Baptist-and-bootlegger coalition formed on the domestic level. Environmentalists, pushing for scrubbers, joined forces with eastern coal producers, who produced fuel higher in sulfur than their counterparts in the western United States. If regulations mandated lower sulfur output rather than specifying the process by which it could be achieved, then eastern utilities would be likely to meet standards by purchasing low-sulfur coal from the West. With scrubbers, however, eastern utilities could use higher-sulfur coal from the East. Eastern coal producers realized they could not avoid regulation altogether, so they joined with environmental actors to achieve the regulation that would be least harmful to their interests.[12]

The 1977 amendments,[13] in addition to mandating scrubbers, extended the deadline for U.S. states to meet the primary air standards set in the 1970 version of the act. These amendments also directed the EPA to regulate, with

[8] Sec.110 (c).

[9] Sec.111, 112.

[10] Sec.202–34.

[11] Bruce Ackerman and William T. Hassler, *Clean Coal/Dirty Air* (New Haven: Yale University Press, 1981), p. 27.

[12] Ackerman and Hassler, *Clean Coal/Dirty Air,* p. 31; David Vogel, *Trading Up: Consumer and Environmental Regulation in a Global Economy* (Cambridge, Mass.: Harvard University Press, 1995), p. 20.

[13] P.L. 95–95; 91 Stat. 726; 42 U.S.C. 7502 (a) (1) (b). It was also amended, though less substantively, in 1971, 1973, 1974, and 1976 to give waivers for the application of motor vehicle emissions standards.

a margin for safety, those substances that could cause serious health problems in ambient air.[14] The act also allows for the creation of new regulations when emissions contribute to air pollution in a way "that may reasonably be expected to constitute a significant danger to health, safety, or welfare of persons," including those in another country.[15]

Industry responded to the increasingly strict air pollution requirements by organizing its lobbying efforts under the Clean Air Working Group. These actors pushed largely for cost-benefit analysis of regulations and for protection of economic growth despite regulations. Environmentalists formed an analogous group, the National Clean Air Coalition, to push for stronger legislation.[16] The processes that created the most recent version of the act are complicated and beyond the scope of this book, but it is worth pointing out that the competitive disadvantage of domestic regulation came up in the debate. Congress had to make decisions about the extent to which becoming a leader in global environmental regulation would disadvantage U.S. firms with respect to their international competitors.[17]

The Clean Air Act Amendments of 1990 are the most significant of the changes to the regulations so far.[18] New were a more comprehensive regulatory program to control acid rain, stricter emission standards for motor vehicles, technology-based controls for toxic air emissions, stricter regulation of ozone-depleting substances, a tradable permit program for air pollution, stricter enforcement, and harsher penalties for noncompliance.

The centerpiece of clean air regulations throughout the history of the Clean Air Act are the National Ambient Air Quality Standards (NAAQS). These have been established for sulfur dioxide ($SO_2$), nitrogen oxides ($NO_x$), carbon monoxide (CO), ozone ($O_3$),[19] lead (Pb), and particulate matter. The EPA sets these standards nationally; the states determine how to implement

[14]  Sec.112, 1977.

[15]  Sec.115.

[16]  Gary C. Bryner, *Blue Skies, Green Politics: The Clean Air Act of 1990 and Its Implementation,* 2nd ed. (Washington, D.C.: CQ Press, 1995), pp. 103–4.

[17]  Ibid., p. 109.

[18]  P.L. 101–549; 104 Stat. 2399 (1990).

[19]  At ground-level ozone is a pollutant, whereas in the stratosphere it provides a shield from harmful ultraviolet rays of the sun. Its primary precursors are volatile organic compouds (VOCs) and it is largely through the regulation of these that ambient ozone is controlled.

them. There are some specific regulations states are required to follow, though the extent to which a given area meets the NAAQS determines how stringently the regulations have to be applied.

From its inception the Clean Air Act has more closely regulated new sources of pollution than existing ones, on the assumption that retrofitting existing sources of pollution would be less cost-effective than requiring abatement technology for new pollution sources. New factories have a greater degree of flexibility about design and where they locate than do existing factories,[20] though the political power of industries with existing plants also probably played a role in establishing regulatory strategies.[21] The EPA's emission standards must reflect "the degree of emission reduction achievable" through best available technology, taking into consideration other requirements relating to air quality, health, environmental protection, and energy efficiency.[22] The New Source Pollution Standards (NSPS) are applied to facilities that began construction (or were significantly reconstructed or modified) after the date that these standards were first proposed. The 1990 Clean Air Act amendments, however, require operating permits for all air pollution sources that specify the level of pollutants they are allowed to emit.

Mobile sources of air pollution, most importantly automobiles, are regulated through both equipment standards and fuel requirements. Since 1975 all automobiles have been required to have catalytic converters. Increasingly strict standards for emissions over the history of the Clean Air Act have led to additional technological changes in the construction of automobiles. Standards for the gasoline used in automobiles first required removing its lead content, then decreasing its volatility (the speed at which it evaporates into the air), and requiring additives to it that lower the level of emissions of certain regulated chemicals. The amount of sulfur in diesel fuel is also limited.

Emission of sulfur dioxide and of nitrogen oxides is also regulated, both as criteria pollutants and for the purpose of preventing acid rain. Coal-burning electricity-generating plants are the major source of sulfur dioxide. Initial regulation of this substance required use of control technology

---

[20] See Senate, Report no. 91–1196, 91st Cong., 2nd sess., 1970, pp. 15–16.

[21] See Kenneth Oye and James H. Maxwell, "Self Interest and Environmental Management," *Journal of Theoretical Politics* 6, no. 4 (1995): 607–8.

[22] Sec.111; 40 C.F.R. part 60.

to remove sulfur from emissions. The 1990 amendments set a ceiling on sulfur dioxide emissions. Various industries can determine how to implement those limits through a system of permits (which allow the emission of a certain amount of $SO_2$) that can be traded.

One way that emissions standards for sulfur dioxide have been met is through the use of low-sulfur coal. Although this change is particularly devastating for the midwestern and eastern suppliers of high-sulfur coal, the cost to the coal industry would not suggest they desire internationalization of regulations, since, if anything, producers of high-sulfur coal would rather that potential foreign markets for their goods not be removed through additional regulation. Instead, the costs borne by industries are the costs of switching coal and the resulting transportation costs, since 85 percent of low-sulfur coal is produced in the West and two-thirds of coal consumption is located in the East.[23]

The other major way in which industry meets $SO_2$ regulations is through the purchase of flue gas desulfurizers, or scrubbers, that remove much of the sulfur from plant emissions. A scrubber can add 10 to 20 percent to the installation cost of a power plant, and operation and maintenance costs also increase.[24]

Sulfur dioxide production in the United States is due largely to electricity generation, which is responsible for two-thirds of all sulfur dioxide generated.[25] Although deregulation of power generation is starting to result in international competition for energy provision, electricity generation at the time that internationalization was a possibility was not an internationally competive industry. Coal is also burned for residential heating and for restaurants, two other uses that do not involve international industrial competition. Other industries that produce sulfur oxides, albeit in lower quantities, include paper mills, smelters, steel mills, and refineries.[26]

Automobiles are the most important sources of $NO_x$, CO, and many volatile organic compounds (VOCs) and they account for more than 50

[23] Michael L. McKinney and Robert M. Schoch, *Environmental Science: Systems and Solutions* (Minneapolis/St. Paul: West, 1996), p. 470.

[24] Ibid., p. 471.

[25] Bryner, *Blue Skies, Green Politics*, p. 68–69. Note that this is the statistic *after* more than twenty years of Clean Air Act regulations. In 1970 the percentage was higher.

[26] Ibid., p. 68.

percent of all air pollution in the United States overall. They are also responsible for more than half of the toxic air pollutants specifically.[27] There are two major ways in which motor vehicles are required to meet air pollution standards in the United States. The first is through the use of reformulated gasoline, a fuel created, for example, by mixing oxygen-rich liquids into the gasoline so that when it combusts it forms $CO_2$, rather than the more harmful CO. In this process VOCs are also oxidized more effectively. A similar though less frequently used alternative relies upon the use of different fuels altogether, such as those mixed with alcohol. Big oil companies fought this aspect (included as a requirement in the 1990 amendments for areas that do not meet carbon monoxide and ozone standards), in part, because producing these fuels requires costly new processes and refining equipment.[28] The second way motor vehicles are required to meet air pollution standards involves postcombustion equipment that removes pollutants before they are released in exhaust. Of these the most widely used is the catalytic converter, which is now standard on automobiles. The 1990 amendments required that automobiles meet stricter emission standards, many of which are to be met through equipment as well.

Fuel and auto manufacturers would certainly want foreign-produced fuel or automobiles sold in the United States to have to meet the same standards they do. It is not clear that to advocate for this would count as pushing for internationalization. Rather, the legislation is written to require these standards to be met for gasoline or automobiles in the United States, regardless of where they are produced, and does not say anything about the requirement to meet these standards in general when not operating in the United States. It is possible that there would be some small advantage to U.S. refiners or auto manufacturers if all foreign competitors had to meet these standards when producing for other markets as well. This element is unlikely to play a large role in pressure for internationalization, however. In the first place, the United States exports inconsequentially small levels of gasoline, so it is not competing with foreign companies outside the United States. Likewise, even though the United States sells automobiles

[27] George Hager, "The 'White House Effect' Opens a Long-Locked Political Door," *Congressional Quarterly Weekly Report*, 20 January 1990, p. 143; Bryner, *Blue Skies, Green Politics*, p. 77.

[28] Bryner, *Blue Skies, Green Politics*, p. 162.

abroad, when they are sold elsewhere they no longer have to meet U.S. domestic standards and can be produced differently for foreign markets if doing so is less costly. We would not, therefore, expect industry pressure for internationalization of motor vehicle regulations from U.S. air pollution regulation.

Toxic air pollution is also regulated by the Clean Air Act. The regulated substances in this category until recently were asbestos, beryllium, mercury, vinyl chloride, arsenic, radionuclides, and benzene.[29] Initially the Clean Air Act required the EPA to list hazardous air pollutants and to establish emission standards for them. But the agency managed to establish limits for only the seven substances listed above and was in the process of establishing limits for coke oven emissions when the act was amended again. The 1990 Amendments listed a total of 189 regulated substances, and requires the EPA to set standards for them by 2000. Polluters are required to use the "maximum available control technology" (MACT) for the substances on this list. The legislation also gives additional time to comply with regulations to industries that reduce emissions by 90 percent before the MACT standards go into effect.

The major nonautomobile sources of releases of toxic air pollutants are the chemical industries themselves.[30] These industries, like coal producers, have no incentive to internationalize. If anything, they will lose potential foreign markets if others regulate the way the United States has. Many other users of toxic air pollutants do not compete abroad. Dry cleaners, for instance, do not compete with foreign unregulated industry, as people rarely send their clothes across borders to be cleaned.

The final major category of air pollution regulations is that pertaining to the prevention of ozone depletion. Clean Air Act regulations have been used to regulate the use of ozone-depleting substances since 1977, when amendments in that year delegated to the EPA the power to regulate substances "anticipated to affect the stratosphere" in a way that could "endanger public health or welfare."[31] The EPA prohibited, in 1978, the use of

[29] David Durenberger, "Air Toxics: The Problem," *EPA Journal* (January/February 1991): 30.

[30] Environmental Protection Agency, *Toxic Release Inventory* (Washington, D.C.: EPA, 1994), pp. 206–208; see also Bryner, *Blue Skies, Green Politics,* p. 77.

[31] 42 U.S.C. 7457 (b).

chlorofluorocarbons (CFCs) in nonessential aerosols.[32] When the Montreal Protocol was signed in 1987, the EPA passed additional regulations to implement the specific requirements of that treaty that were not already present in U.S. law. The 1990 Clean Air Act amendments restrict production, use, emissions, and disposal of ozone-depleting substances, though they go farther in some respects than is required by the Montreal Protocol and its amendments.[33]

Ozone-depleting-substance regulations impose costs on a number of different industries. The primary producers of CFCs had to bear initial costs of losing some of their market, owing to domestic regulation, and had to spend research funds to develop substitutes if they were to continue to dominate the market. Large users of regulated ozone-depleting substances would also be harmed if the United States continued its regulation of these chemicals, since the few substitutes that were available were much more expensive.

## Internationalization of Air Pollution Regulations

Most aspects of Clean Air Act regulations have some requirements that need to be met by other countries, since goods produced for consumption in the United States are generally required to meet U.S. standards wherever they originate. Automobiles produced in Japan must have catalytic converters in order to be sold in the United States, motor vehicles in general must be able to run on the type of fuel they are required to use in the United States, and products sold there are not allowed to emit toxic fumes. This aspect of regulation is almost always already international in any kind of domestic regulation in any country, however. Trade treaties recognize as legitimate the right of states to exclude products for internal consumption that do not meet their domestic safety standards. This type of internationalization does not concern us, since there is no necessary attempt to convince other states to adopt a regulation for all the products they produce and does not involve the way they make them. U.S. domestic indus-

---

[32]  43 *FR* 11301; 43 *FR* 11318.

[33]  It thus, in fact, provides a push to work for further internationalization of these regulations within the Montreal Protocol process.

tries would benefit if these factors were truly internationalized, however. If foreign auto makers or energy producers, for instance, had to meet U.S. standards, regardless of where they made or sold their products, it would likely cost them more and would be of competitive advantage to U.S. firms.

The only real U.S. efforts at internationalizing air pollution regulations relate to those restricting the use of ozone-depleting substances. To some extent regulations preventing acid rain have also been internationalized, although that case represents a struggle between the United States and Canada about whether and which regulations would be adopted bilaterally, and the United States dragged its feet on negotiating. Interestingly, both of these regulations have been internationalized mainly through international agreement rather than solely through economic pressure, although the possibility of economic threats was also discussed by Congress in working toward internationalization of these regulations.

### Unsuccessful Internationalization Efforts—General

There were some not-quite-successful attempts to create legislation to internationalize elements of Clean Air Act regulations. Unlike those unsuccessful efforts in other issue areas to create a domestic process for pushing regulations internationally, these efforts very nearly succeeded in creating domestic processes and are worthy of discussion in an effort to understand internationalization attempts that did succeed domestically.

In the Senate process of amending the Clean Air Act in 1990 Slade Gorton (R-Wash.) proposed imposing a tariff on "any product imported into the United States that has been subject to processing, or manufactured from a process, which does not comply with the air quality standards of the Clean Air Act." The amendment was defeated, narrowly, by a vote of 52 to 47. Senator Frank Lautenberg (D-N.J.) offered a related amendment to a different piece of legislation, after Gorton's amendment was defeated. The amendment, to Section 301 of the Trade Act of 1974, would have considered "a failure to establish . . . effective pollution abatement and control standards to protect the air" as an unreasonable trade practice that would qualify for U.S. trade restrictions. David Boren (D-Okla.) proposed a third Senate bill the following year, that would have defined foreign pollution as a production subsidy, and allowed countervailing duties against states whose pollution policies "did not meet domestic American production standards."[34] None of these pieces of proposed legislation had particularly strong environmental support, however. In terms of air pollution priorities, this type of internationalization was low on the agenda of environmental groups. These bills were not connected to any particular type of environmental harm from air pollution; environmental organizations did not lobby in favor of them in Congress; none of these bills passed.

## Acid Rain

Some of the sources of acid rain have been regulated in the United States since the passage of the Clean Air Act. Substances like $SO_2$ and $NO_x$, the main contributors to long-range, transboundary air pollution, have been regulated under the NAAQS for their ambient air pollution effects. The United States measured these pollutants, however, by examining whether they were present on the ground. This form of measurement encouraged the building of tall smokestacks to carry pollutants away from the source. To some extent, then, this regulation increased the transboundary aspect of air pollution from the United States.[35]

In the late 1970s the United States and Canada funded national studies on transboundary air pollution and established a joint scientific Research Consultation Group to exchange information on acid rain. In 1980 the United States and Canada signed a Memorandum of Intent, agreeing to take interim measures to combat transboundary air pollution, and scheduled negotiations to create a formal agreement. Domestic industry groups in the United States, however, prevented the negotiations from moving forward.[36]

There are important environmental reasons for international control of the processes that produce acid rain. Given that emissions in one country cause environmental damage in another, acid rain–producing pollution is a problem unlikely to be solved unilaterally. U.S. environmental organiza-

---

[34] David Vogel *Trading Up,* pp. 209–10.

[35] National Research Council, *Atmosphere-Biosphere Interactions: Toward a Better Understanding of the Consequences of Fossil Fuel Combustion* (Washington, D.C.: National Academy Press, 1981), p. 142.

[36] Vicki L. Golich and Terry Forrest Young, "United States-Canadian Negotiations for Acid Rain Controls," Case 452, *Pew Case Studies in International Affairs,* Institute for the Study of Diplomacy, Pew Case Studies Center, Georgetown University, 1993, p. 4.

tions lobbied Congress and spoke out publicly on the need for negotiation with Canada for an agreement. When it became clear that Congress would not authorize serious negotiations, environmental groups tried other approaches. Organizations like the Natural Resources Defense Council joined northeastern states in suing the EPA for allowing Midwestern utilities to cause acid rain in Canada. The court initially ruled against the EPA, though the ruling was overturned on appeal.[37]

The United States and Canada eventually signed an executive agreement, the United States-Canada Agreement on Air Quality, in March 1991. The agreement did require Canada to take some measures that were analogous to those the United States had taken under the Clean Air Act. In addition, it requires both states to cut acid rain emissions in half by 2000 and to put permanent emission caps "at an acceptable level." It also involves promises to cooperate on developing technology and on regulatory principles, and to monitor emissions.

The push for this agreement, however, came primarily from Canada, so it is not obvious that this case should be considered internationalization of U.S. domestic air pollution regulations. If anything, it was probably more accurately considered to be internationalization of Canada's domestic regulation. After the Memorandum of Intent, Canada increased its level of regulation of the substances and processes that cause acid rain, even though the United States resisted taking steps toward cooperation during the Reagan presidency. The Environment Minister of Canada commented that "we will proceed independently of the United States in the hope that they will join us."[38] The United States eventually did, under the Agreement on Air Quality.

### Ozone Depletion

The United States did decide to internationalize regulations pertaining to ozone depletion. As outlined above, aspects of the Clean Air Act (as well as some other separate pieces of legislation) required U.S. industry, under 1977 Clean Air Act legislation, to phase out use of CFCs in nonessential

[37] Rochelle L. Stanfield, "Environmentalists Try the Backdoor Approach to Tackling Acid Rain," *National Journal* 17, no. 2 (1985): 2365ff. (Lexis/Nexis).

[38] Eliot Marshall, "Canada Goes It Alone on Acid Rain Controls," *Science* 226 (21 December 1984): 1275.

aerosols, beginning in 1978. Domestic industries that either made or relied on CFCs challenged the regulations as soon as they were proposed. DuPont, the largest U.S. chemical company and the one responsible for most CFC production, argued that unilaterally limiting U.S. use of CFCs would "harm the American economy without yielding any measurable environmental benefits."[39] Utah Representative Dan Marriott went so far as to ask President Reagan to suspend enforcement of EPA CFC regulations.[40] The chemical industry formed a lobbying group, the Alliance for Responsible CFC Policy, and worked to combat proposals for further regulation of ozone-depleting substances. Among other tactics, it drafted legislation to be introduced into Congress to limit the EPA's ability to regulate ozone-depleting substances.[41]

At the same time, scientific and environmental organizations were lobbying for stricter standards on ozone-depleting chemicals. New studies appeared regularly, indicating the worsening condition of the ozone layer and the increasing certainty that anthropogenic chemicals were responsible for it. Environmental NGOs argued for further regulation. The Natural Resources Defense Council sued the EPA in U.S. District Court to require it to regulate emissions of CFCs more stringently, which, the group argued, was required under the Clean Air Act. They also supported holding foreign CFC producers to U.S. standards.[42]

The United States worked from the time of initial domestic regulation to gain international protection of the ozone layer. The United States hosted in 1977 the International Conference on the Ozone Layer, which produced the World Plan of Action on the Ozone Layer, and was an important participant in the 1985 Vienna Convention for the Protection of the Ozone Layer, which supported the principle of ozone layer protection but did not require actual reduction in emissions of ozone-depleting substances.

[39] "Chemical Giant Raps EPA Action," *U.P.I.*, 16 April 1980 (Lexis/Nexis).

[40] "Regional News—Utah," *U.P.I.*, 8 February 1991 (Lexis/Nexis).

[41] "FYI," *PR Newswire*, 17 February 1981 (Lexis/Nexis); R.C.C. "Congress Debates Depletion of Ozone in the Stratosphere," *Christian Science Monitor*, 14 October 1981: 19 (Lexis/Nexis).

[42] Marjorie Sun, "Lawsuit Seeks a Cap on Fluorocarbon Production," *Science* 226 (14 December 1984): 1297; Statement of David Doniger, Natural Resources Defense Council in Senate, *Stratospheric Ozone Depletion and Chlorofluorocarbons*, 100th Congress, 1st sess., 12, 13, and 14 May 1987, pp. 321–23.

After Vienna it became clear that there was support for lower limits on U.S. production and consumption of ozone-depleting substances, and that these limits would likely be undertaken whether or not there was an international agreement to do so. At this point CFC industry actors made a calculation similar to the one made by the eastern coal producers in negotiation of the Clean Air Act. If they were going to be regulated, they would rather be regulated on their own terms. In 1986 individual chemical companies, like DuPont, and the industry's collective body, the Alliance for Responsible CFC Policy, came out in favor of international regulation on ozone-depleting substances. Richard Barnett, of the Alliance, noted that "controls on CFC's solely in the United States would produce little, if any, environmental benefit and could significantly harm our competitive position in the world economy."[43]

The United States took a leadership position in the 1987 negotiation of the Montreal Protocol to the Vienna Convention. Throughout the process the United States pushed for deep cuts in the use of ozone-depleting substances. During the international negotiations, Congress worked on domestic legislation that would internationalize U.S. policy regardless of the outcome of the international negotiations. Several bills introduced into Congress in early 1987 would have unilaterally prohibited U.S. imports of ozone-depleting substances, or products containing or made with ozone-depleting substances, if the exporting countries did not have national ozone regulations equivalent to those of the United States.[44] These bills were intended to be used if states were not willing to negotiate international controls on ozone-depleting substances. Press coverage of the legislation was pervasive during the protocol negotiations. Moreover, the U.S. negotiator, Richard Benedick, made sure that other states knew of this possibility during the negotiations.[45] When the protocol passed, the legislation became unnecessary and was dropped.

[43] Richard Barnett, "Business Forum: Saving the Earth's Ozone Layer; The U.S. Can't Do the Job All Alone," *New York Times,* 16 November 1986: Section 3, p. 2 (Lexis/Nexis); Richard Elliot Benedick, *Ozone Diplomacy: New Directions in Safeguarding the Planet* (Cambridge, Mass.: Harvard University Press, 1991), p. 31.

[44] 100th Congress, 1st Session, S.579, S.571, and H.R.2036.

[45] Karen T. Litfin, "Framing Science: Precautionary Discourse and the Ozone Treaties," *Millennium* 24, no. 2: 265.

The ozone case deserves a postscript, however, to indicate that those who initiate the internationalization process cannot necessarily determine its ultimate implementation. Large-scale U.S. producers and consumers of ozone-depleting substances joined forces with environmental organizations in an effort to ensure that theirs would not be the only industries required to take action on this issue. They hoped that their competitors would have to limit use of ozone-depleting substances as well. The internationalization process they created, however, took on a life of its own. Eventually it resulted in the complete phaseout of certain ozone-depleting substances that were initially only regulated. In the long run, though, international regulation did more to protect some industry interests than unilateral domestic regulation would have. The press for internationalization, then, may begin with a domestic environmental interest and may be joined by a domestic industry interest. Overall, however, environmental exigencies may prevent industry actors from achieving their internationalization goals in the form they intended.

### Addendum: Automobile Regulations

Although automobile-related regulations were not internationalized per se, some of the economic incentives for internationalization did lead to action on the part of the United States that bore some similarities to internationalization efforts and showed the power of the automobile industry. In the first place, technology and emission standards on U.S. cars were also placed on foreign vehicles. In fact, some of these standards were constructed in such a way that foreign producers were subject to a competitive disadvantage in the U.S. market. One of these, the Corporate Average Fuel Economy (or CAFE) program,[46] requires fuel efficiency standards, but measures them across an entire fleet of vehicles sold in the United States from a particular manufacturer. Whereas this regulation appeared nondiscriminatory, it had the effect of imposing penalties on automobile manufacturers that produce primarily luxury cars, which are less fuel-efficient. Ninety-nine percent of fines for exceeding CAFE standards were imposed on European auto manufacturers.[47] A GATT panel declared the regulation to be contrary to international trade law.

---

[46] Not strictly part of the Clean Air Act, but related.

[47] Paul Stanton Kibel, "GATT Fouls the Air," *Recorder,* 14 November 1994: 8 (Lexis/Nexis).

In the second place, foreign producers of gasoline for the U.S. market were required to reformulate their gasoline, once U.S. producers were required to. But again, the standards that were applied discriminated against foreign producers. Clean Air Act regulations from 1990 requiring reformulated gasoline (RFG) stated that conventional gasoline emissions could not exceed 1990 emissions levels (and had to be reduced by 15 percent in the most polluted areas), to make sure that pollutants from RFG were not put into the conventional gasoline. U.S. refiners were allowed to use 1990 measures from their own refineries, but foreign refiners were required to use a baseline of average quality of gasoline in 1990, which could make it harder for them to meet the standard. Venezuela and Brazil took the case to the newly formed World Trade Organization dispute-settlement process; the panel ruled against the United States, which has now offered to allow more regulatory options to foreign fuel producers.[48]

Because in neither of these cases did the United States attempt to convince foreign industry to meet U.S. Clean Air standards for non-U.S. markets, these instances do not represent attempts at internationalization. They are significant, however, in showing the political power of the U.S. automobile industry in working not only to level the regulatory playing field but to tilt it to the advantage of the United States. If there had been a convincing environmental partner in this case, it is likely that these regulations would have been subject to the type of internationalization attempts examined in this volume.

## Air Pollution Conclusions

If we consider ozone layer protection the only true case of internationalization among air pollution regulations, there are too many, rather than too few, candidates to explain internationalization of that regulation. As indicated already, a number of different observers have explained internationalization of ozone depletion regulations as resulting from various elements of both environmental and economic externalities. For this reason internationalization of ozone depletion regulations has been treated

[48] "EPA Changes for RFG Imports," *Oil and Gas Journal,* 12 May 1997 (Lexis/Nexis); Alan Tonelson and Lori Wallach, "We Told You So: The WTO's First Trade Decision Vindicates the Warnings of Critics," *Washington Post,* 5 May 1996: C4 (Lexis/Nexis).

throughout this volume as a tool through which to develop and present explanations for internationalization more generally. Ozone layer protection should ultimately be examined along with the cases in the other issue areas to help differentiate among different explanations. We can, however, look at the aspects of the Clean Air Act that have not been internationalized to help understand the conditions under which internationalization is mostly likely.

Neither the environmental externalities or economic externalities hypotheses alone can explain the pattern of internationalization observed. A number of regulations that have important international environmental and economic effects have not been pushed internationally by the United States. The one regulation that has been internationalized in this set is the one that benefits both industry actors and environmentalists. Those regulations that have not been internationalized were supported only by one group and not the other.

**Environmental Externalities**
There are clearly environmental aspects to efforts to internationalize clean air regulations. Certainly ozone depletion was an environmental problem that could not be addressed without international action, and it was something that could produce environmental harm worldwide. A general concern for environmental protection was put forth by some of the actors pushing other attempts at internationalization. Senator Gorton spoke in favor of his proposed Clean Air Act amendment by saying that "we live in one world . . . [and] should do everything we can to encourage policies which are similar to our own in the rest of the world, whether it has a direct impact on air quality in the United States or not."[49]

Given the fact that environmental damage from air pollution has been a concern for so long, one would not be surprised to find versions of the environmental explanation playing a role in determining which regulations the United States attempts to internationalize. On the premise of the primacy of an environmental externalities explanation for internationalization, one would expect to find internationalization of the portions of the Clean Air Act that address pollutants that cause the most serious environmental harm,

[49] Vogel, *Trading Up*, p. 210.

by relating to particularly dangerous or pervasive pollutants, or by addressing those pollutants that cross borders.

The Clean Air Act regulates activities that lead to different types of environmental harm and, unlike the straightforward process of determining which species is rarer than another, it is not always obvious how to determine how much environmental harm they cause. One way to calculate the environmental harm from air pollution is to determine how toxic, or dangerous, the pollutants released are in and of themselves. Another way to determine environmental harm from air pollution is to prioritize those pollutants that are emitted in the greatest quantities. Chemicals that are toxic and somewhat widely used, such as lead, should be of large concern. Also important should be those that are less toxic but widely emitted, carbon monoxide primary among them. Particulate matter might be important here, because it is a significant pollutant in industrialized countries and also in the less developed world. Those pollutants that are (or have been) somewhat widely used that persist in the atmosphere, like CFCs, should also be targets of internationalization on the basis of the environmental harm they cause. It might also be important to take into consideration the persistence in the air of the emitted pollutants. These range from 1 to 4 days for oxides of sulfur and nitrogen to 75 to 100 years for CFCs. In the middle are CO, with an atmospheric half-life of 75 days, and arsenic and lead, with half-lives of up to 30 days.[50] Those with longer half-lives may persist in the environment in greater quantities than those, emitted at higher levels, that decompose more quickly.

If internationalization of regulations is attempted for those pollutants that cause the most harm from international sources, we would expect the extent to which air pollution crosses borders to determine the likelihood of internationalization regulations. Provisions for dealing with pollution relating to global commons issues—like those pertaining to ozone-depleting substances—should be particularly important targets of internationalization efforts, since the United States could not protect its own territory if others did not regulate these substances as well. Non-commons transboundary problems, like acid rain, might also be important targets of internationalization. Although there is no inherent reason that this formulation would need to

---

[50] William P. Cunningham and Barbara Woodworth Saigo, *Environmental Science: A Global Concern* (Dubuque, Iowa: William C. Brown), pp. 387–91.

concern the direction of the pollution (since any issue in which environmental harm crosses borders requires action by all polluters regardless of the direction in which it flows), it is possible that the direction of the pollution would matter as well, with the United States having the greatest incentive to internationalize those regulations pertaining to transboundary pollution that harms it.

Direct environmental harm does not account for the attempted internationalization we see. There are substances regulated under the Clean Air Act, such as asbestos, mercury, and radionuclides, that are harmful to humans at small doses, and there is no evidence of any serious attempt to internationalize regulations pertaining to these substances. Moreover, the CFCs and related substances that cause ozone depletion are themselves quite harmless directly to humans; it is precisely for this reason that the substances were so widely adopted. In terms of overall harm, ozone-depleting substances certainly cause serious environmental damage, but it is not clear that their harm is significantly more problematic than the harm caused by other air pollutants like sulfur dioxide and carbon monoxide. In addition, emissions of particulate matter provide the highest percentage of air pollutants in developing countries and thus are responsible for environmental harm in large portions of the world. Particulate air pollution from anthropogenic sources totals approximately a hundred million metric tons per year.[51] If environmental harm were at the source of internationalization efforts, we would expect regulations pertaining to particulates to be a high priority.

The attempted internationalization of regulations pertaining to ozone depletion is consistent with the concerns about the transboundary nature of air pollution, but efforts to internationalize regulations related to sulfur dioxide, nitrogen oxides, and volatile organic compounds, responsible for acid rain, would be too. The fact that the United States was a source of the environmental harm crossing the border into Canada, as well as a recipient of acidified air from Canada,[52] makes the situation less clear in terms of the environmental harm experienced by the United States, since it causes as well as suffers environmental damage. But if transboundary environmental

---

[51] Ibid., p. 390.

[52] Roderick W. Shaw, "Acid-Rain Negotiations in North America and Europe: A Study in Contrast," in *International Environmental Negotiations,* ed. Gunner Sjostedt, (Newbery Park: Sage Publications, 1993).

concerns were the main determinant of internationalization priorities, we would have predicted greater efforts internationally on behalf of acid rain.

**Economic Externalities**

There is clearly an economic element in the attempts to internationalize air pollution regulations, particularly concern for the competitive disadvantage that air pollution regulations cause for U.S. domestic actors. The format of the proposed legislation to authorize economic restrictions on products made in ways that harm the ozone layer can give some insight into the logic behind internationalization policies. Like the amendment offered to the 1990 Clean Air Act to restrict trade, this legislation did not pass. It is nevertheless interesting that the sanctions considered for ozone-depleting substances were based, not on where the harm was coming from, but on where the competition was. Sanctions would have prevented entry of goods produced in a way that does not meet U.S. standards for prevention of ozone depletion. Since the level and area of ozone depletion are not related to the location of the use of ozone-depleting substances, concern for the environment would not require that these restrictions be targeted on any one area. If anything, they would be most appropriately targeted against those states that damage the ozone layer the most, through greatest production or use of ozone-depleting substances, rather than against manufacturers who happen to make goods for trade with the United States. Clearly, then, environmental considerations were not the main ones behind this proposed legislation.

The same thing is expressed even more clearly in the various attempts to legislate to address air pollution regulations in other countries. The legislation specifically refers to the trade advantages ("unfair subsidies") gained by those who do not have the same level of air pollution regulation as the United States. Industry actors tried to play the economic competitiveness card in opposing various stages of increasing domestic regulation under the Clean Air Act. The president of the National Coal Association pointed out that "reasonably priced electricity over the past decade has meant economic growth and an improvement in the competitiveness of U.S. manufactured goods, both at home and abroad." He noted that "without reasonably priced energy and readily available electricity, the accomplishments of the past cannot continue." He characterized the acid rain provisions of the act as "counterproductive to our Nation's competitive performance potential in the international marketplace."[53] That the Clean Air Act was made stricter domestically over time, however, and that none of these proposed trade restrictions for air pollution internationalization passed, indicate that international competitiveness alone was not sufficient to result in the internationalization of domestic air pollution regulations.

Were one to assert the primacy of an economic externalities explanation for internationalization, one would expect internationalization of those portions of the Clean Air Act that impose a cost on internationally competing domestic industries, or of those portions that, when internationalized, provide opportunities for profit by industries that have already adapted to the domestic regulations. We might expect, therefore, to find attempts at internationalization of policies that have little or no transboundary environmental effects, as long as they impose a cost on those industries that compete across borders. One would not be surprised to find a strong economic component present in decisions about whether to internationalize U.S. air pollution regulations. There are large costs to concentrated (and therefore politically influential) industries for meeting Clean Air Act standards, and U.S. technology has been an important tool for addressing air pollution. If economic externalities are at the root of attempted internationalization of air pollution regulations, it should not matter whether the pollution itself crosses borders; it is simply the economic effects of domestic regulation that should determine internationalization.

If internationalization is due to the competitive disadvantage of U.S. regulations on U.S. industries, we would expect internationalization of air pollution regulations that impose high costs on U.S. industries that compete abroad. Those that produce things purely for a domestic market in which actors from other states do not compete, such as energy, should not be internationalized. As a logical extension of this argument one could imagine that any internationally competing industry that uses inputs from domestic industries subject to the U.S. air pollution requirements would have an incentive to work for internationalization of domestic air pollution regulations.

[53] Richard L. Lawson, President, National Coal Association, testimony presented on 3 October 1989 in hearings before the Environmental Protection Subcommittee of the Senate Environment and Public Works Committee, as quoted in *Congressional Digest* 69, no. 3 (1989): 87.

Energy might cost more in the United States because of environmental regulations, for example. Industries that have to purchase energy as an input would also bear a higher cost than their foreign competitors. But this level of competitive disadvantage is unlikely to be serious enough (or politically salient enough), when combined with other cost differences that firms across countries bear, to be a clear determinant of internationalization.

If internationalization pressure comes from those industries that would benefit from giving others an incentive to use the technology they have created, we would expect internationalization of regulations for which U.S. actors have a competitive advantage in abatement technology. Several have even noted the presence of this aspect on the domestic level in amending the Clean Air Act. Although not strictly "technological," producers of low-sulfur coal supported strong sulfur emissions regulations in 1977 since they had a competitive advantage in the coal that would likely be used to meet the standards.[54] Actors who produce (or have patented) the technology used to comply with air pollution regulations would benefit from the greater demand created by internationalizing the regulation.

A 1996 Environmental Protection Agency report showed that the costs to U.S. industries in complying with the Clean Air Act have indeed been high. Adjusted for inflation, direct annual costs of compliance with the regulations of the Act ranged from about $20 to $25 billion.[55] There are certainly industries that suffer from the regulations. Attaching a scrubber to a smokestack can cost $100 million.[56] Clean Air Act requirements are blamed for driving production costs up 25 percent in the coal- and steel-mining industries.[57] The EPA report indicated that the total cost to taxpayers and consumers of the CAA regulations between 1970 and 1990 was $436 billion.[58]

[54] Vogel, *Trading Up*, p. 20.

[55] Environmental Protection Agency, *The Benefits and Costs of the Clean Air Act, 1970 to 1990*, Draft. Prepared for the U.S. Congress. October 1996, p. 10. These amounts are adjusted to 1990 dollars.

[56] Stanfield, "Environmentalists Try the Back Door," p. 2365ff.

[57] "Now the Squeeze on Metals," *Business Week*, 2 July 1979: 46 (Lexis/Nexis).

[58] EPA, *Benefits and Costs*; Gerald Karey, "Clean Air Study: Big Costs, but Worth It," *Platt's Oligram News*, 11 June 1996: 4 (Lexis/Nexis).

Unlike the situations in either fisheries or endangered species regulations, in that in air pollution regulations technology has played an important part since their inception. The earliest way to address local air pollution was through technology—building higher smokestacks so the pollution would travel away from the area in which it was created. In air pollution regulations in general the United States has exported more technology than it has imported. In the 1980s the United States exported air pollution control technology annually worth between $88 and $163 million, and imported annually regulatory technology worth between $21 and $97 million.[59] The technology used to meet Clean Air Act regulations on criteria pollutants consists mainly of flue-gas desulfurizers, and parts (such as catalytic converters) for automobiles to control emissions. U.S. industries make and export both these types of technology, giving them an incentive to work for internationalization of these regulations. In neither case, however, are they the only important producers, so the incentive is weaker in these cases than in others.

A greater incentive for internationalization is felt by the major CFC-producing industries in the United States, foremost among them DuPont. The company was the original inventor (with General Motors) of CFCs and was responsible for more than half the world market of the substances. It invested tens of millions of dollars in the late 1980s to researching CFC substitutes.[60] It developed, for commercial use, HCFCs and HFCs. The former of these depletes the ozone layer, though significantly less than CFCs. Both have been widely used as CFC substitutes. DuPont's early action on substitutes and its dominance in the industry gave it a strong advantage to push for internationalization for technological competitive advantage.[61]

But competitive advantage, often given as an explanation for U.S. internationalization of ozone regulations, seems insufficient to bring about

[59] Environmental Protection Agency, *International Trade in Environmental Protection Equipment: An Assessment of Existing Data.* EPA 230-R-93-006 (July 1993), p. 10.

[60] Edward A. Parson, "Protecting the Ozone Layer," in *Institutions for the Earth: Sources of Effective International Environmental Protection,* ed. Peter M. Haas, Robert O. Keohane, and Marc A. Levy (Cambridge, Mass.: MIT Press, 1993), pp. 41, 46.

[61] Oye and Maxwell, "Self-Interest and Environmental Management," pp. 193–201.

internationalization on its own. The failed legislative efforts to internationalize general competitiveness aspects of the Clean Air Act indicate that the broad issue of competitive advantage does not alone sway policymakers.

In particular, the United States had a competitive advantage (especially over Canada) in air pollution control technology. For most of the 1970s and 1980s, Canadian coal plants had not installed scrubbers, and Canadian auto emission standards were weaker than those in the United States.[62] These factors should have provided an economic incentive for industry actors to work for internationalization of acid rain regulations. For instance, Canada has been the largest purchaser of U.S. environmental protection technology in general, accounting for 21 percent of U.S. exports of this equipment.[63] Instead the United States resisted forming an agreement with Canada to reduce acid rain.

The most convincing explanation for this resistance is the realistic fear that any agreement would not only involve internationalizing current regulations, but would require the United States to make deeper cuts in its emission of acidifying pollution than it already had.[64] This resistance shows, however, that competitive advantage (or disadvantage) was not a sufficient concern to push for internationalization of regulations, since Canada would presumably have had to regulate comparatively further than the United States would have. The effects of acid rain are determined both by the direction of prevailing winds and by the ecological sensitivity of the regions onto which it is deposited; the harm felt within the United States comes more from internal than external sources. Half of the acid rain in Canada comes from pollution emitted in the United States. Conversely, 15 percent of the acid rain in the United States comes from Canadian sources. So while the acid rain problem could not be completely solved within the United States without the involvement of Canada, due to the transboundary nature of the problem, the United States is a greater net exporter of acid rain. Examining environmental harm in conjunction with

[62] Golich and Young, "United States-Canadian Negotiations," p. 17.

[63] EPA, *International Trade in Equipment,* p. 15. The specific wording of the proposed amendment, however, leaves no doubt that it stems from economic competitiveness concerns.

[64] This is the same thing that eventually happened with ozone protection regulations on the international level.

transboundary issues begins to explain the lack of internationalization of acid rain provisions.[65]

## Baptists and Bootleggers

If the combined interests of industry and environmental actors are responsible for internationalization, the only air pollution regulations likely to be internationalized are those that offer both environmentalists and industry an incentive. It is not surprising that ozone protection regulations were pushed internationally, since almost any explanation for the internationalization would cover the ozone case. Nevertheless, a Baptist-and-bootlegger explanation would count the ozone case among the most likely candidates for internationalization.

In failed efforts to create domestic policy for internationalization we have further evidence that both environmentalist and industry actors are important to create such a policy. Acid rain policy needed to be addressed internationally for environmental reasons, and it had the support of domestic environmental groups who were even willing to sue the Environmental Protection Agency in pursuit of international goals. But with industry opposed to North American acid rain negotiations, U.S. efforts to take the lead on them failed. Likewise, general legislation proposed in Congress to level the economic playing field for internationally competing industries harmed economically by the Clean Air Act failed when it did not have sufficient environmental justification for proceeding. So although U.S. efforts internationally on behalf of the ozone layer are overdetermined, the political experiences that shaped the overall set of air pollution regulations indicate that both environmental and competitive reasons for internationalization are needed for the U.S. government to make a decision to work to apply U.S. regulations to foreign states.

[65] Ned Helme and Chris Neme, "Acid Rain: The Problem," *EPA Journal* (January/February 1991): 19; Golich and Young, "United States-Canadian Negotiations," p. 4.

# 5

# Ocean Fisheries

Regulation of open-ocean fisheries is an issue that is inherently international. The ocean is a commons, and regulation by one state cannot adequately conserve ocean fisheries resources if other states do not adopt similar regulations. Fish do not recognize the lines in the ocean that have been drawn by states attempting to regulate resources. Most regulation of open-ocean fisheries therefore has at least some international component.

Fisheries are considered here, in part, because they provide a slightly different domestic context for internationalization than do the other issue areas examined. Many of the U.S. regulations borne by U.S. fishers actually originate at the international level, through the activities of multilateral organizations. For example, U.S. tuna fishers in the Atlantic Ocean are bound mostly by rules originally passed by the International Commission for the Conservation of Atlantic Tunas, an international organization operating since 1969, whose regulations become law in the United States. These regulations, international in origin, might therefore initially seem unlikely candidates for the process of U.S. internationalization. Examining U.S. responses to these regulations, therefore, gives us an opportunity to see if, and how, the United States attempts to further internationalize issues that are already regulated on the international level. What would internationalization mean in such a context? It could take the form of pushing the international organization to adopt more stringent requirements to match stricter domestic regulations, or of encouraging more states to accept international regulations than already have, or of ensuring that states that have adopted regulations implement them.

In addition, regulation of fisheries, unlike that of endangered species, is regulation of a renewable resource intended for sustainable use. The goal of fisheries management is to ensure that a population of fish reproduces itself in sufficient numbers so that fishing can continue indefinitely. Fisheries management is therefore much more like forestry than like preservation of an endangered species, and as such represents an important but often overlooked segment of international environmental policy.

Successful regulation of international fisheries is difficult and seldom accomplished.[1] There are often environmental reasons to work to make regulation of international fisheries more successful; there are also economic costs to fishing restrictions as well as to the depletion of fish stocks, so there is likely to be industry concern over the shape of international fisheries regulation. Thus, international fisheries seem to offer conditions we have seen in other issue areas that provide the potential for internationalization, despite the difficulty.

## U.S. Legislation

Unlike endangered species regulations or air pollution regulations, major fisheries regulations are not unified in a single piece of legislation. These rules are passed individually by Congress and listed in the *U.S. Code*. The regulations considered here are all the major regulations on open-ocean fisheries that U.S. fishers are required to uphold, most of which derive from the regulatory actions of international organizations. These regulations are generally promulgated on the basis of a species, of a region, or of both. There are also some domestic regulations that impose or prohibit certain types of fishing gear or require licenses, and more recently there have been catch limits or seasons for some species within two hundred miles of the U.S. coast.

The history of fisheries regulations falls into two somewhat distinct periods. Until the late 1970s fisheries further than three or twelve miles from coastlines were all regulated internationally. Efforts to claim jurisdiction over an area two hundred miles from a state's coastline were initiated by Latin American states in the 1940s. They based their claims on the U.S. Truman Proclamation, of 1945, that referred to jurisdiction over the continental shelf. The general international recognition of Exclusive Economic Zones (EEZs),

---

[1] See, for example, M. J. Peterson, "International Fisheries Management," in Peter M. Haas, Robert O. Keohane, and Marc A. Levy, *Institutions for the Earth* (Cambridge, Mass.: MIT Press, 1993), pp. 249–305; Michael Berrill, *The Plundered Seas: Can the World's Fish Be Saved?* (San Francisco: Sierra Club Books, 1997).

within which states were responsible for fishery conservation, was accomplished by the mid-1970s. The 1982 completion of the third United Nations Conference on the Law of the Sea legitimized the idea of a two-hundred-mile EEZ, and by 1988 ninety-two states had established "Exclusive Fisheries Zones" and another seventy-two had declared EEZs.[2] The United States initially resisted this move toward international acceptance of EEZs. The Magnuson Fishery Conservation and Management Act, enacted in 1976, extended U.S. fishery jurisdiction to two hundred miles (though the United States had not yet declared an EEZ) and created Regional Fishery Management Councils to make regional policies to regulate fishing, and to oversee foreign fishing within the two hundred-mile area. The Act excludes "highly migratory species of fish," defined as "tuna," from U.S. jurisdiction in the zone.[3]

The major international fisheries commissions from which the United States gets domestic fisheries regulations are as follows. The Inter-American Tropical Tuna Commission (IATTC) regulates tuna and tuna-like species in the Eastern Tropical Pacific Ocean.[4] U.S. tuna fishers are also regulated under an analogous and connected agreement, codified in U.S. law as the Eastern Pacific Tuna Licensing Act, designed to continue regulations between the United States and Costa Rica once the IATTC experienced difficulties regulating tuna catch.[5] The International Commission for the Conservation of Atlantic Tunas (ICCAT) regulates tuna and tuna-like species in the Atlantic Ocean. It began operations in 1969. The North Pacific Fisheries Commission (NPFC) regulates all the major commercial fisheries of the northeast Pacific Ocean. It has regulated these fisheries since 1952 and was amended in 1978 to incorporate EEZs into its regulations.[6] The International Commission for Northwest Atlantic Fisheries

---

[2] See James C. F. Wang, *Handbook on Ocean Politics and Law* (New York: Greenwood Press, 1992), pp. 44, 109. The designation of the mid-seventies as a turning point in fisheries regulation is a somewhat arbitrary but nevertheless accurate reflection of the time that the notion of EEZs influenced international fisheries management. Nineteen seventy-six to seventy-seven is the dividing line used by M. J. Peterson in her analysis of international fisheries regulation, "International Fisheries Managment," p. 250.

[3] 16 U.S.C. Sec.1813, 1802.

[4] 16 U.S.C. 951.

[5] 16 U.S.C. 972 (1984).

[6] Signed on 9 May 1952, amended on 25 April 1978; implemented in U.S. law at 16 U.S.C. 1021.

(ICNAF) regulated the main commercial fisheries off the coasts of the United States, Canada, Greenland, and eastern Iceland, beginning in 1950 and ending in 1979,[7] but was replaced the following year in part by the North Atlantic Fisheries Organization (NAFO). This latter organization, however, regulates only fish stocks that straddle the two hundred-mile EEZ off Canada. Established in 1982, the North Atlantic Salmon Conservation Organization (NASCO) inherited ICNAF's responsibility for regulating salmon in the North Atlantic.[8] The North East Atlantic Fisheries Commission (NEAFC) regulates the main commercial species in that region. It began operations in 1959 and was revised in 1979 to take EEZs into account and regulate those species only beyond two hundred miles from coastlines. The Committee for the East Central Atlantic Fisheries (CECAF) has regulated fisheries in the northern Atlantic off the coast of Africa and the Gulf of Guinea since 1969. The Commission for the Conservation of Antarctic Marine Living Resources (CCAMLR) regulates the krill and finfish fisheries of the southern ocean, beginning in 1982.[9] The Western Central Atlantic Fishery Commission (WCAFC) regulates fisheries in the Caribbean and the Gulf of Mexico. The Indo-Pacific Fisheries Commission (IPFC) regulates fisheries off Southeast Asia and the east coast of India. The Indian Ocean Fisheries Commission (IOFC) regulates stocks in that region.

There are also several United States-Canada fishery organizations that regulate the actions of United States fishers. Halibut is regulated in part by the commission under the 1953 Convention between the United States of America and Canada for the Preservation of the Halibut Fishery of the Northern Pacific Ocean and Bering Sea.[10] The 1985 Treaty between the Government of the United States of America and the Government of Canada Concerning Pacific salmon regulates salmon that spawn in U.S. and Canadian rivers.[11]

[7] Implemented in domestic law in 1950; 16 U.S.C. 981–91; repealed by P.L. 95-6, section 4, in 1977; 91 Stat. 16.

[8] 16 U.S.C. 3601.

[9] The treaty entered into force for the United States in 1982 and is implemented in domestic law at 16 U.S.C. 2431; T.I.A.S. 10240; 33 U.S.T. 3476.

[10] 16 U.S.C. 773; the convention was amended in 1979.

[11] 16 U.S.C. 3631.

Some fishing regulations that U.S. fishers are bound by are considered in the section on endangered species regulation, because although they regulate fishing activity, they do so in the name of protecting endangered species. These include regulations on shrimp fishing to protect sea turtles and regulations on yellowfin tuna fishing to protect dolphins. In addition whaling regulations could be considered in this chapter because whales have been regulated internationally as a fishery. U.S.-specific regulations, however, regulate whales as endangered species, so they are considered in chapter 3.

## Internationalization of Fisheries Regulations

The United States passed legislation to allow internationalization of several general aspects of fisheries regulation; regulations relating to specific species of tuna and to salmon; and regulations against fishing with driftnets. Other equipment regulations for fisheries that the United States attempted to internationalize include restrictions on the use of purse-seine nets for catching Pacific tuna, and a requirement for using Turtle-Excluder Devices when catching shrimp. Both of these regulations are discussed in chapter 3. Many of the regulations for which the United States worked to gain broader international acceptance had their origins in international regulations, many of which the United States was instrumental in establishing. In addressing fishery conservation, then, "internationalization" involves attempts to gain acceptance of and compliance with these regulations by states that did not initially regulate.

The main U.S. efforts at internationalization (or further internationalization) of fishery regulations relate to tuna and to the prevention of the use of driftnets. Each is the subject of at least three different legislative efforts to gain acceptance of U.S. regulations by fishers of other states. Protection of salmon was important as well. It was the subject of the Pelly Amendment, which formed the basis of many later internationalization attempts, and conservation of salmon was one of the concerns behind the banning of driftnet fishing. The only other legislation for internationalization of fishery regulations, in the U.S. Antarctic Marine Living Resources Convention Act of 1984, addressed conservation of krill, an important environmental resource that was expected to have a substantial economic benefit. It might have been an issue of competitive disadvantage had the krill fishery turned out to be overfished. Instead, however, the collapse of the Soviet Union and the chaos into which the economy of the region was thrown removed the main krill-fishing state from the region and thus prevented competitiveness issues from arising.[12]

### General Internationalization of Fisheries Policy

The first piece of relevant legislation in the U.S. effort to work for international regulation of fisheries was part of the 1962 U.S. Trade Expansion Act.[13] This act authorizes the president to raise tariffs up to 50 percent above the tariff rate from 1 July 1934 on fish from countries that refuse to negotiate in good faith for fishery conservation agreements. It was not internationalization of the type examined here because it does not address specific regulations to which U.S. fishers are subject, but it was an early indication that the United States was willing to undertake trade restrictions on behalf of international protection of fisheries.

### Salmon—The Pelly Amendment

U.S. fishers under the International Convention for the Northwest Atlantic Fisheries (ICNAF) were subject to restrictions on the amount of salmon they could catch. Although fishers would benefit collectively over the long run from regulation of the overall level of salmon that could be caught, individual short-run interests would be hurt by limiting their catch. Moreover, the long-term advantages of regulation would be negated if not all fishers participated in regulatory arrangements. When ICNAF first set quotas on salmon catches in the early 1970s, Denmark, Norway, and West Germany used the objections procedure under the convention, which meant that they were not bound by the quota.[14] U.S. fishers reacted angrily, since they were the main fishers whose salmon catches were restricted; in other words, U.S. industry actors experienced competitive disadvantage. At the same time, the experience provided a lesson for conservation organizations since it demonstrated that simply belonging to an international organization for

---

[12] Christopher C. Joyner, "Managing Common-Pool Marine Living Resources: Lessons from the Southern Ocean Experience," in *Anarchy and the Environment: The International Relations of Common Pool Resources,* ed. J. Samuel Barkin and George Shambaugh (Albany: SUNY Press, 1999), pp. 70–96.

[13] P.L. 87-794; 76 Stat. 872.

[14] Legislative History, p. 2412; from House Report 92-468.

natural resource conservation was not sufficient to guarantee that member states would uphold the conservation agreements reached.

These groups lobbied Congress in support of what became the most influential legislation for holding foreign actors to regulations that bound their U.S. counterparts. Since the context was salmon, industry organizations like the International Atlantic Salmon Foundation and the Committee on the Atlantic Salmon Emergency spoke out about the harm that would befall U.S. salmon fishers if they were bound to regulations that their competitors were not. General environmental organizations, such as the National Wildlife Federation, also weighed in on the importance of being able to hold foreign fishers to international standards, both in the context of salmon, and more generally.[15]

The law that passed in 1971 came to be known as the Pelly Amendment to the Fisherman's Protective Act.[16] It allows for unilateral U.S. sanctions "when the Secretary of Commerce finds that nationals of a foreign country, directly or indirectly, are conducting fishing operations in a manner or under circumstances which diminish the effectiveness of an international fishery conservation program." In such an instance, the secretary certifies the situation to the president, who then has the option to "prohibit the bringing or importation into the United States of fish products." Within sixty days of certification the president must report to Congress on any action taken relating to the certification and, if he does not impose sanctions, the reason that no action was taken.[17] This general legislation has been the model for many of the regulatory measures specific to particular species. Congress directly addressed the question of what counted as an "international fishery conservation agreement" for the purpose of engaging U.S. sanctions. The State Department witnesses suggested that an agreement be "widely held," but the House Merchant Marine and Fisheries Committee stated that it "believes rather that all multi-lateral conservation agreements, be there 3 or 15 signatory nations, stand on equal footing," as long as the agreement is in force.[18] As this criterion was applied, however, "in force" did not mean

[15]  *U.S. Code Congressional and Administrative News*, p. 2409, 1971, from House Report 92-468.

[16]  P.L. 92-219; the Fisherman's Protective Act was originally of 1967.

[17]  P.L. 92-219 Sec. 8 (a) and (b); 22 U.S.C. §1978.

[18]  Legislative History, p. 2416; from House Report 92-468.

that the obligation in question had to be legally in force for the state against which sanctions were applied. The United States was thus able to threaten trade restrictions for states that did not uphold international standards the United States supported, whether or not these target states had indicated a willingness to undertake these regulations.

## Tuna I

Under U.S. law tuna fishers are subject to regulations by two main international bodies, as indicated previously: the Inter-American Tropical Tuna Commission (IATTC) and the International Commission for Conservation of Atlantic Tunas (ICCAT). The IATTC, created in 1949, regulates tuna fishing in the eastern Pacific Ocean. The commission initially required tuna fishers to provide catch statistics, and beginning in 1966 designated areas, seasons, minimum-catch lengths, and an overall quota for tuna.[19] Quotas were implemented only until 1979. After 1979 the states in the commission were unable to agree on or implement quotas, in part due to the increasing demands of most states to regulate tuna fishing within their own EEZs, a move the United States opposed. The remaining members of the commission still conduct studies and attempt to protect marine mammals in the course of tuna fishing.

In the Atlantic Ocean U.S. tuna fishers are subject to the regulations of the International Commission for the Conservation of Atlantic Tunas (ICCAT), a body created by convention in 1966. The objective of the convention is "to maintain tuna populations at levels which permit the maximum sustainable catch for food and other purposes."[20] The area of concern for the commission is the Atlantic Ocean and adjacent seas. Twenty-two states are party to the agreement, including the United States, Canada, Japan, France, and a number of developing states. Regulations adopted by the commission include minimum catch-size regulations for tropical tuna, bluefin tuna, and swordfish; fishing mortality limits for bluefin tuna and swordfish; and some quotas. For purposes of management (and because of a theory that two distinct stocks of bluefin tuna populate the

---

[19] IATTC, *Annual Report 1967* (La Jolla, Calif.: 1968).

[20] Alberto Szekely and Barbara Kwiatkowska, "Marine Living Resources," in *The Effectiveness of International Environmental Agreements: A Survey of Existing Legal Instruments*, ed. Peter H. Sand (Cambridge, UK: Grotius Publications), p. 276.

Atlantic Ocean) the ocean has been divided at the 45th parallel into east and west regions. The ICCAT has imposed catch quotas for West Atlantic bluefin tuna. Unlike the Inter-American Tropical Tuna Commission, the ICCAT divides its quota among the major tuna-fishing states in the region: the United States, Canada, and Japan.

Tuna fishing in the South Pacific is regulated by the Treaty on Fisheries between the Governments of Certain Pacific Island States and the Government of the United States of America,[21] which regulates tuna fishing in the EEZs of the states party to the agreement.

The creation of the IATTC made tuna fishers wary. On the one hand, they would benefit from a regime to collect information and make collective conservation decisions to protect the tuna stock. On the other hand, fishers feared they would be bound when their counterparts in other states were not, thereby harming them economically and diminishing any environmental benefits from the agreement. Those concerned more generally with conservation noted that the creation of some mechanism to hold foreign fishers to the same obligations undertaken by those in the United States would help conservation more broadly, since the first goal of the IATTC was simply the collection and analysis of information.[22]

The Tuna Conventions Act of 1950 authorizes the United States "to prohibit from entry from any country all yellowfin tuna taken by vessels of such countries under circumstances which tend to diminish the effectiveness of the conservation recommendations of the [Inter-American Tropical Tuna] Commission."[23] It also allows for an embargo on other types of tuna that are "under investigation by the Commission and which were taken in the regulatory area."[24] The U.S. legislation recognizes that "some of the countries are not parties to the Convention and, therefore, have no applicable treaty obligations to fulfill" but that "the achievement of the conservation

[21] This agreement entered into force for the United States on 15 June 1988, and is implemented domestically at 16 U.S.C. 973.

[22] *U.S. Code Congressional and Administrative News,* 1971, pp. 3612–13, from Senate Report No. 2094, 18 July 1950.

[23] 40 *FR* 48159; 16 U.S.C. 955 (c). Imports of tuna from the country in question may be allowed if accompanied by a certificate of eligibility, allowed in certain circumstances outlined in 50 CFR Sec. 281.7.

[24] 40 *FR* 42230.

objectives . . . is dependent upon international cooperative efforts to implement the Commission's recommendations."[25] The 1975 Atlantic Tunas Convention Act contains provisions nearly identical to those for the Pacific process, for ensuring international conservation of Atlantic tuna.[26]

## Tuna II—EEZs

Another element of U.S. regulations on tuna involved the issue of who would be allowed to put restrictions on catching tuna. Historically, tuna, which swim long distances, have been regulated, not by any one state, but by international agreement. That practice came under attack with the growing acceptance of states' abilities to regulate resources within two hundred miles of their shores, an acceptance the U.S. government initially resisted. Tuna fishers opposed this principle as well, since it would limit their access to tuna, most of which could be found at any particular time within two hundred miles of some coast.

More complex was the position of conservation organizations. On the one hand, the U.S. government position avoiding domestic consideration of tuna-fishing regulation was transparently designed to protect U.S. tuna-fishing interests, particularly once all other fishery regulations within two hundred miles were accepted. On the other hand, it was likely that international regulation *was* the best way to conserve fishery resources, and for that reason some environmental organizations supported a move to require international regulation.

The United States therefore attempted to internationalize its resistance to EEZs, particularly as they relate to tuna. Since tuna is a highly migratory species and does not remain within any one state's EEZ, the United States maintained for a period of time that tuna could only be regulated internationally. In this case the United States attempted to prevent individual states from regulating tuna within their EEZs. It can be questioned whether this phenomenon is actually "internationalization," but the United States presented the issue as such, looking for acceptance of the principle that tuna would only be regulated internationally. This effort first began with the Fisherman's Protective Act (FPA), which is the act to which the Pelly Amendment was later attached. The FPA was enacted in 1954 to respond

[25]  50 CFR §281.2 (d).
[26]  16 U.S.C. 971.

to the seizure of U.S. fishing boats within two hundred miles of the coasts of Latin American states that had declared two hundred-mile fishery zones.[27] It provides for compensation to fishers for boats seized, revenue foregone, and legal fees for fishers arrested for violation of fishing laws of foreign governments.[28] If this amount was not recovered from the government in question, it was deducted from the U.S. foreign aid supplied to that state. This deduction automatically occurred, unless the president or secretary of state determined that it would harm the U.S. national interest to make the deduction.[29] It has been suggested that this legislation was enacted to encourage fishers to continue to fish in these zones during the period when the United States was resisting the creation of customary international law that would recognize two hundred-mile fishery zones. Once the United States had its own two hundred-mile zone, the legislation was used for the same purpose for tuna, which the United States excluded from its jurisdiction. The FPA was amended in 1967 and 1978, as well as by the 1971 Pelly Amendment.

The Magnuson Fishery Conservation and Management Act, which excluded tuna from the species that states should be allowed to regulate individually, also had provisions to discourage states from setting tuna-fishing limits outside of international organizations. In what can essentially be seen as an effort to protect the access of U.S. tuna fishers to the EEZs of other states, the legislation calls for the secretary of the treasury to declare an embargo on the importation of fish products from any state that does not allow U.S. vessels to fish for tuna within its EEZ.[30] The first step in this process is that the secretary of state certifies that such lack of access occurs and is either "in violation of an applicable fishery agreement . . . without authorization under an agreement between the United States and such a nation; or as a consequence of a claim of jurisdiction which is not recognized by the United States."[31] The secretary of the treasury is then required "to prohibit the importation into the United States (1) of all fish and fish

---

[27] N. Peter Rasmussen, "The Tuna War: Fishery Jurisdiction in International Law," *University of Illinois Law Review* 1981, no. 3: 762.

[28] 22 U.S.C. 1974ff.

[29] 22 U.S.C. 1978.

[30] 16 U.S.C. 1825.

[31] 16 U.S.C. 1825.

products from the fishery involved . . . and (2) upon recommendation of the Secretary of State, such other fish or fish products, from any fishery of the foreign nation concerned, which the Secretary of State finds to be appropriate to carry out the purposes of this section."[32] In other words, the United States has the authority to embargo tuna or whatever U.S. fishers were trying to fish for, and the secretary of state would have the discretion to embargo other fish products, too. Although this measure may seem only to be about access to the resources of other states, the United States justified it as a tuna conservation measure.

### Antarctic Fishery Resources

The southern ocean contains a wide variety of fishery resources. Owing in part to its geographic distance from the major fishing states, the southern ocean retained undepleted resources longer than other productive fishing areas. Regulation of Antarctic resources in general is done under the rubric of the Antarctic Treaty System. In 1980 the parties to the treaty negotiated the Convention for the Conservation of Antarctic Marine Living Resources (CCAMLR). This agreement attempts to regulate marine resources through an ecosystem approach, taking account of the effects of the depletion of different marine resources on the ecosystem as a whole. Catch quotas are nevertheless imposed on krill and finfish. At the time of the negotiation of CCAMLR the major fishers in the region were from Japan, Poland, West Germany, and the Soviet Union, fishing mostly for krill and finfish.[33] U.S. fishers did not take many resources from the southern ocean, but were considering future resource exploitation, with the encouragement of the U.S. government.[34]

When Congress considered legislation to implement the convention, it considered the possibility of an expanding U.S. fishery in the region and the fear that regulations would keep U.S. fishers from the level of fishing enjoyed by other states. It passed the U.S. Antarctic Marine Living Resources Convention Act of 1984,[35] which bans possession, sale, import, or export,

---

[32]  16 U.S.C. 1825.

[33]  Heidi Mathiesen, "Antarctica: Cutting up a Frozen Pie," *Christian Science Monitor,* 16 April 1982: 22 (Lexis/Nexis).

[34]  Robert Reinhold, "As Others Seek to Exploit Antarctic, U.S. Takes the Scientific Approach," *New York Times,* 21 December 1981: D13 (Lexis/Nexis).

[35]  16 U.S.C. 2435 (3) (1984).

of Antarctic marine living resources (or their parts) taken in violation of the Convention on the Conservation of Antarctic Marine Living Resources.

## Driftnets

For a number of reasons relating to overharvesting of fishery resources and endangering marine mammals, U.S. fishers have been prohibited from fishing with driftnets within the United States' EEZ and on the open ocean. Beginning in 1990, the Marine Mammal Protection Act mandated a zero marine mammal mortality rate when using driftnets, essentially abolishing any remaining U.S. driftnetting industry. In addition, on 22 December 1989 the UN General Assembly passed Resolution 44/225 on "large-scale pelagic driftnet fishing and its impact on the living marine resources of the world's oceans and seas." This resolution recommends a moratorium on "all large-scale pelagic driftnet fishing on the high seas by 30 June 1992" in the absence of "effective conservation and management measures." It calls, as well, for cessation of driftnet fishing in the South Pacific by 1 July 1991, and "immediate cessation of further expansion of large-scale pelagic driftnet fishing on the high seas of the North Pacific and all the other high seas outside the Pacific Ocean."[36] The resolution itself was introduced by the United States, at the urging of some environmentally concerned members of Congress,[37] and the United States followed it up by making fishing with driftnets in that area illegal.

The banning of driftnets had long been the rallying cry of environmental organizations and others concerned with sustainable harvesting of ocean resources. Driftnets entangle birds and marine mammals, causing their deaths, and catch large numbers of fish as bycatch. During the 1980s, the National Marine Fisheries Service estimated that driftnets killed up to fifty thousand fur seals.[38] Moreover, nets that are lost may drift for years, killing fish and other creatures indiscriminately.

Some environmental organizations tried confrontational approaches to halt foreign driftnet fishing. The Sea Shepherd Conservation Society sent a 169-foot vessel out in the summer of 1987 to prevent Asian vessels from

---

[36] UN General Assembly Resolution 44/225.

[37] *U.S. Code Congressional and Administrative News,* 1990, p. 6282.

[38] Philip Shabecoff, "Huge Drifting Nets Raise Fears for an Ocean's Fish," *New York Times,* 21 March 1989: C1.

using their driftnets. The Japanese fleets cut short the fishing season that summer.[39] Others tried legislative approaches, with representatives from groups such as the Center for Environmental Education and the International Marine Mammal Project testifying before Congress for an end to foreign driftnet fishing.[40]

U.S. fishers in the Pacific Ocean fished with driftnets through the 1980s until domestic restrictions grew tighter and new ones were imposed. The major users of driftnets in the Pacific Ocean were Japan, South Korea, and Taiwan. Though fishing primarily for squid, they would catch millions of pounds of salmon as bycatch, further depleting an already-disappearing stock.[41] In addition, under United States-Canadian agreements on Pacific salmon, catches of this fish species are strictly limited and allowed on the basis of where the stock originates. Asian fishing vessels were thus catching salmon designated for North American fishers.[42] Because of that, U.S. salmon fishers joined the call for regulation on foreign driftnets. Even the Sport Fishing Institute added its voice to the opposition to driftnets, documenting Taiwanese driftnet fishers dumping hundreds of illegally caught salmon overboard.[43]

The call to end the use of driftnets by foreign fishers was therefore not a difficult political battle within the United States Congress.[44] The impressive array of support for efforts to end foreign driftnet fishing eventually resulted in the passage of three pieces of legislation to allow the United States to work for an end to driftnet fishing internationally. The Driftnet Impact Monitoring Assessment and Control Act of 1987 directs the Department of Commerce to monitor and report on environmental effects of the use of driftnets by foreign nations within and beyond the U.S. Fishery Conservation Zone and to negotiate cooperative agreements to monitor and enforce

---

[39] "Sea Shepherd's [sic] Claims Victory over Driftnet Fleets," *U.P.I.*, 21 July 1987 (Lexis/Nexis).

[40] Elmer W. Lammi, "Washington News," *U.P.I.*, 30 April 1987 (Lexis/Nexis).

[41] "Commerce Department Says South Korea and Taiwan Violate Driftnet Act," *U.P.I.*, 29 June 1989 (Lexis/Nexis).

[42] Shabecoff, "Drifting Nets," p. C1.

[43] Gene Mueller, "U.S. Authorities Are Too Soft on Foreign Fish Netters," *Washington Times*, 15 August 1989: D8 (Lexis/Nexis).

[44] Donald E. DeKeiffer, "Pyrrhic Victory in 'Driftnet War,'" *Journal of Commerce*, 24 July 1989: 8A (Lexis/Nexis).

driftnet fishing rules. Under this act, the Pelly Amendment was expanded to allow imposition of U.S. sanctions on fish products from states that refused to enter into agreements to ensure "effective enforcement of law, regulations and agreements" about the use of driftnets, or those that refused to ban the use of driftnets by 29 June 1989.[45]

Section 206 of the Fishery Conservation Amendments of 1990 on Large-Scale Driftnet Fishing[46] allowed Pelly certification for states that "conduct, or authorize their nationals to conduct, large-scale driftnet fishing beyond the exclusive economic zone of any nation in a manner that diminishes the effectiveness of or is inconsistent with any international agreement governing large-scale driftnet fishing to which the United States is a party or otherwise subscribes."[47]

Finally, the High Seas Driftnet Fisheries Enforcement Act[48] of 1992 amends the Pelly Amendment to allow the United States to restrict imports of fish and fish products from states "engaged in driftnet fishing or related trade." The act requires that the secretary of commerce, after 10 January 1993, identify any country whose nationals "are conducting large-scale driftnet fishing beyond the exclusive economic zone of any nation." Within thirty days after such identification the president must consult with that state for the purpose of gaining an agreement to terminate large-scale high seas driftnet fishing by that state. If within ninety days the consultation does not result in a successful agreement, the president is to direct the secretary of the treasury "to prohibit the importation into the United States of fish and fish products and sport fishing equipment" from the state in question.[49]

## Fisheries Conclusions

The U.S. thus, at the urging of a wide variety of domestic actors, worked to internationalize a variety of fisheries regulations relating primarily to tuna,

---

[45] Title IV of the United States-Japan Fishery Agreement Approval Act of 1987, P.L. 100-220, 101 Stat. 1477–78, 16 U.S.C. 1822.

[46] P.L. 101–627, Title I, section 107; 104 Stat. 4443.

[47] Ibid.

[48] P.L. 102-582; 106 Stat. 4900.

[49] P.L. 102-582, Section 101.

salmon, and driftnets. Moved by less obvious motives, it worked to gain international acceptance of Antarctic fishery regulation. Most of these regulations have both environmental and economic incentives for internationalization; what determines whether the United States pushes a fishery regulation internationally?

## Environmental Externalities

Environmental externalities provide several different types of incentives for internationalization of fishery regulations, though they alone do not explain the pattern of regulations the United States pushes internationally. If environmental reasons were the main source for U.S. internationalization efforts with respect to fisheries issues, we would see internationalization of regulations pertaining to fishery problems that are either the most environmentally pressing or the most international in scope. Those fisheries that are the most depleted, such as those involving fish species listed in the U.S. Endangered Species Act or in the appendices of the Convention on International Trade in Endangered Species (CITES), or to those that have the lowest population numbers relative to preexploitation levels, would be most protected internationally, as would fish species that migrate long distances or across multiple jurisdictions.

The fisheries protected by the rules the United States attempted to internationalize are all severely depleted. Bluefin tuna, for instance, are so overfished that they have been declared an endangered species under the U.S. Endangered Species Act, and have been proposed for listing under CITES. It is certainly among the most depleted of all fisheries. The Northern Bluefin stock in 1996 comprised less than 25 percent of 1970 levels. The ICCAT estimates that between the mid-1980s and the mid-1990s stocks older than eight years fell by 87 percent. The Southern Bluefin tuna stock has also reached a record low.[50] Five species of Pacific salmon are listed as endangered or threatened under the Endangered Species Act. Though eleven to fifteen million salmon once spawned annually in the Columbia River system in the western United States, the level is now down to three million, most of which come from

[50] Ted Williams, "The Last Bluefin Hunt," *Audubon,* July-August 1992: 14–20; World Wildlife Fund, "The Large Pelagic Fishes," http://www.panda.org/research/fishfile2/fish42.htm, date visited: 26 June 1996.

salmon hatcheries.[51] Certainly, the fish species the United States has worked hardest to protect internationally are environmentally deserving of that attention.

But an examination of fishery regulations that the United States did not push internationally shows that the level of depletion cannot be the only factor in internationalization. Pacific halibut and herring are among the rarest of the commercially exploited fish,[52] and yet the United States did not attempt to internationalize regulations relating to these species. The United States has also not worked to protect sturgeon internationally, despite species of sturgeon being listed in the ESA and their being the only fish regulated in the United States listed on either Appendix of CITES.[53] Other ocean fishery stocks regulated for U.S. fishers and seen as severely depleted include cod, haddock, hake, plaice, herring, halibut, and Pacific Ocean perch.[54] For none of these was there any serious effort to pursue greater international protection. The collapse of the international cod fishery, for instance, is legendary,[55] yet, there have been no real efforts to push domestic restrictions internationally.

The United States has also worked to improve international protection for fish species that are particularly transboundary in behavior. At first glance it might seem that all ocean fishery regulations would be of an equally

[51] Fish and Wildlife Service, "Fact Sheet on Threatened and Endangered Species," http://www.fws.gov/bio-salm.html, 26 June 1996; "The Tragedy of the Oceans," *Economist,* 19 March 1994: 21.

[52] "The Tragedy of the Oceans," p. 22. These fish are so depleted that the seasons during which they can be fished are shorter than two days each year. In 1994 the season to catch herring roe was a mere forty minutes long.

[53] Fish and Wildlife Service, "Fact Sheet on Threatened and Endangered Species," http://www.fws.gov/bio-salm.html, date visited: 26 June 1996; 50 C.F.R. 17.11, pp. 62–65. Other species listed here are freshwater fishes. See also C.F.R. Title 50, chapter I, subchapter B, subpart C, "Appendices I, II, and III to the Convention on International Trade in Endangered Species of Wild Fauna and Flora," (50 C.F.R. 23.23) p., 178; http://www.fws.gov/r9dia/public/sturgfs.html, date visited: 10 September 1998. Baltic Sturgeon are listed on Appendix I.

[54] Carl Safina, "The World's Imperiled Fish," *Scientific American* 273 (November 1995): 46–53; Michael Parfit, "Diminishing Returns: Exploiting the Ocean's Bounty," *National Geographic* (November 1995): 2–37; "The Tragedy of the Oceans," pp. 21–24.

[55] Mark Kurlansky, *Cod: A Biography of the Fish That Changed the World* (Toronto: A.A. Knopf, 1997).

transboundary character, since the oceans can be considered an international commons. Several aspects of the transboundary nature of fishery regulation can differ, however. First, there is variation on how far or how quickly fish travel out of national control. Second, since the advent of two hundred-mile Exclusive Economic Zones (EEZs), states have some territorial control over fish. Since many species of fish either swim out to sea farther than two hundred miles or swim in and out of the EEZs of different states, fish species may vary in the extent to which they are under the control of one or more states.

However one measures it, the most transboundary fish species is tuna. Bluefin tuna can swim at speeds of fifty miles per hour and travel from continent to continent across the open ocean.[56] Yellowfin tuna also travel from the Pacific coast of the United States to South America. Other types of tuna, such as bigeye, skipjack, and albacore, also travel long distances. Salmon spawn in inland rivers and begin their lives in the territory of states, and then swim out to sea in and out of various national jurisdictions in the ocean, and thus can also be seen as highly transboundary. The species in the Antarctic Ocean, the subject of internationalization efforts, are much less migratory, although they exist only outside of territorial boundaries and are in that way among the most international of resources. The fish species the United States has worked hardest to protect internationally are therefore quite transboundary. Other fish species that are highly migratory have been less well protected internationally by U.S. efforts, however. These would include mackerel, cod, flounder, plaice, hake, and herring.[57]

We might also look to other environmental considerations to determine which fish species are the most important to protect. Krill, which was the subject of the push to internationalize Antarctic fishery regulations, is seen as an environmentally central species, since it is the base of the aquatic food chain.[58] All in all, there are a number of important environmental reasons to explain the fishery regulations the United States works to internationalize. But a simple application of environmental exigencies to U.S. regulations would not result in the pattern of internationalization we see with respect to fisheries.

[56] Williams, "Last Bluefin Hunt," p. 14.

[57] "The Tragedy of the Oceans," p. 24.

[58] Joyner, "Managing Common-Pool Marine Sources," pp. 70–96.

## Economic Externalities

Similarly, although we can see the effect of economic externalities on U.S. efforts to internationalize fishery regulations, these causes do not alone explain the efforts the United States makes internationally on behalf of ocean fisheries. An explanation of U.S. efforts based solely on economic externalities would predict the extent and cost of regulation for U.S. fishers to be the most important determinant of the internationalization of fisheries regulations. We might expect to see internationalization of those regulations that impose a domestic cost on U.S. fishers relative to their competitors. Or, internationalization might come for fishery regulations that have some technological aspect to the regulation, for which U.S. industries have already created the technology they could market abroad.

U.S. domestic actors have indeed suffered competitively from many of the fishery regulations the U.S. attempts to internationalize. Many domestic fishery regulations impose costs on U.S. fishers that do not have to be borne by foreign fishers of the same stocks, due either to a cost for fishing equipment or to decreased productivity relative to others. Even when fishing is regulated by an international body, those who have to uphold costly rules when others do not will suffer competitively. It is in this latter context that most U.S. efforts at internationalization fall. In a number of international agreements U.S. fishers are bound by regulations their competitors are not. Fishers of tuna species have been particularly prone to this difficulty. Three-quarters of the states that fish for bluefin tuna are not bound by ICCAT.[59] Likewise, at any given time, a number of yellowfin-fishing states have not been bound by IATTC regulations,[60] or by regulations governing driftnet fishing.

The resistance to EEZs and the accompanying insistence that tuna conservation could be regulated only internationally provide the clearest evidence of the importance of economic externalities in U.S. policy. Despite the environmental rhetoric, the disadvantage to U.S. fishers of losing access to a lucrative fish stock was clearly the incentive behind the U.S. policy of insisting that tuna could not be regulated by states.

---

[59] Williams, "Last Bluefin Hunt," p. 19.

[60] Inter-American Tropical Tuna Commission, *Annual Reports, 1950 through 1980* (LaJolla, Calif.: IATTC, 1952–81).

The importance of the fisheries in question for U.S. fishers, and the severity of the restrictions may also play a role in the policy. Salmon and tuna are the largest fisheries in which U.S. fishers participate. The United States comprises more than half the world market for tuna.[61] An inability to use long driftnets hurts U.S. salmon fishers competing with Asian fishers, since driftnet fishing is the most efficient way to harvest salmon. We might wonder why other economically lucrative species, like sturgeon, are not further protected, but the major U.S. efforts at internationalization do relate to those species U.S. fishers lose most economically in protecting.

Nevertheless, species that are much less economically central, and from the regulation of which the United States does not suffer great consequences, are the subject of regulations the United States works to internationalize. In particular, the push to internationalize the krill and finfish regulations in the southern ocean cannot be explained by purely competitiveness concerns, since U.S. participation in this fishery is relatively small.

In addition, it is clear that the potential economic advantages of internationalization, due to increased markets for technology, do not create economic incentives to internationalize fishery regulations. Technology is frequently blamed for the depletion of the fisheries because it has made fishing more efficient and has allowed fishers to stay out on the oceans longer and travel farther and faster.[62] But in these instances the technology has been responsible for overfishing; rarely is it used as a solution. There is little support for the influence of competitive advantage from technology in the push to internationalize fishery regulations, in part because there are few technological solutions to the fishery problems embodied in the regulations.

### Baptists and Bootleggers

Although economic incentives to internationalize fishery regulations account for most of the attempts at internationalization that have been made, bringing in the environmental incentives and examining the groups of actors pushing both types of agendas can provide a more complete explanation of those attempts. If "Baptists and bootleggers" are behind attempts

---

[61] Dorothy J. Black, "International Trade v. Environmental Protection: The Case of the U.S. Embargo on Mexican Tuna," *Law and Policy in International Business* 24, no. 1 (1992) (Lexis/Nexis).

[62] Parfit, "Diminishing Returns," pp. 18–19; Safina, "World's Imperiled Fish," pp. 48–49; Peterson, "International Fisheries Management," p. 250.

at internationalization, we would expect to see internationalization only of regulations that address concerns of industry actors and environmentalists. Tuna regulations appear in both these categories, as do salmon regulations. Equipment restrictions, particularly involving driftnets, are also of concern to both groups. More importantly, the fishery regulations the United States has pushed internationally are the ones that have engaged domestic groups representing both environmental and economic reasons for international regulation.

On the environment side, it is surprising to see nonindustry actors pushing hard for increased regulation. Fisheries do not often engage the concern of environmental or even scientific groups not directly responsible for their regulation, but interestingly enough regulation of salmon, and to a greater extent of tuna (particularly of bluefin tuna in the Atlantic Ocean) have had important backing of influential environmental groups.

Internationalization attempts at driftnet regulation show the clearest conjunction of interests. There are two main reasons to be concerned about driftnet fishing. The first is that the efficiency of this method of fishing allows many fish to be caught without much difficulty, and leads to the potential for species depletion. The main species at issue is salmon (of U.S. origin). The coalition of actors behind the passage of the original Pelly Amendment was particularly concerned about the loss of salmon that U.S. fishers considered to be proprietary to the United States. The other species important to U.S. industry caught in driftnets is albacore tuna, which related to earlier tuna conservation sanctions. U.S. catches of albacore tuna also declined during this period. Even when foreign fishers were fishing for species (such as squid) U.S. fishers were not seeking, their actions posed an economic threat to U.S. fishers of salmon and tuna. This concern underpins the call for the first set of driftnet sanctions.

The second concern about driftnets, however, is that they entangle and kill marine mammals when they are used for fishing. This concern motivated regulations under the Marine Mammal Protection Act. As a result, U.S. fishers using driftnets had to use costly measures to avoid killing marine mammals while their competitors did not. Because of the marine mammal regulations, the U.S. (salmon and tuna) fishing industry had an added incentive to want to internationalize the domestic regulation. In both cases environmentalists, hoping to preserve more marine mammals, joined industry in this push. Legislation to support internationalization of driftnet regulations was supported by environmental groups concerned about marine mammals, and by domestic salmon fishers who could not use driftnets themselves and who suffered if their competitors did use them.

Other evidence that internationalization of the driftnet regulations involved an industry-environment coalition can be seen in the content of the threatened sanctions. The High-Seas Driftnet Fisheries Enforcement Act, pushed in part by sport-fishing organizations that cared about salmon, included import restrictions on sport-fishing equipment from those states that did not end driftnet fishing. The interests of the coalition members influenced the composition of the threatened sanctions.

Likewise, the two Tuna Conventions Acts were also supported by a mix of industry actors and environmentalists. There was domestic opposition from the U.S. tuna fishing industry to the amendments to the Tuna Conventions Act that gave the National Marine Fisheries Service (NMFS) the authority to put IATTC decisions into place. The U.S. tuna industry feared that it would be disadvantaged by having to abide by international regulations that not all tuna fishers would be held to, since the membership in the IATTC was small.[63] Similarly, tuna fishers in the Atlantic (some of whom were Pacific tuna fishers who would fish for tuna in the Atlantic during the closed season in the Pacific) wanted to ensure that all fishers in the region were regulated to the same extent that they were. And the environmental concerns about tuna, either to regulate it for the preservation of dolphins, or to preserve the large and anthropomorphized bluefin tuna, contributed to industry efforts to internationalize these regulations.

The two cases that fit this model less clearly are the internationalization of Antarctic fishery regulations and the U.S. resistance to regulating tuna within EEZs. The former seems due entirely to environmental concerns and the latter to economic ones. EEZs may to some extent be a special case. It is the one set of regulations considered here in which the United States is trying to resist policy by other states rather than to impose new U.S. policy on them. Moreover, it represents aspects of high politics in a way most environmental cases do not. Not only was the United States upholding the

---

[63] The Convention was originally signed by the United States and Costa Rica; at various times France, Japan, Nicaragua, Panama, Vanuatu, Canada, Ecuador, and Mexico joined, but the latter three later withdrew.

idea of freedom of the seas, but it was responding to actual capture of U.S.-flagged vessels. Importantly, those who favored the policy felt the need to frame it in terms of an environmental goal and to bring in token support of environmentalists. This support was not entirely misguided, since tuna regulation by Latin American states within their EEZs almost universally failed to protect the stocks. Despite the importance of economic considerations to this policy, the environmental elements were not entirely window dressing.

The puzzle of why some deserving fishery regulations have not been pursued internationally becomes clearer when examining the conjunction of economic and environmental interests that underpin internationalization attempts. Without the marine mammal incentives that driftnet and tuna regulations provide, environmental groups have generally not shown much involvement in fishery regulation, at least until the depletion of a stock becomes so severe it cannot be ignored. And unlike the sleek, large, and fast bluefin tuna that have recently attracted the attention of environmental activists, fish like sturgeon and cod have been unable to draw the attention of general conservation organizations. The ability of these organizations to draw funding and participation for broader environmental protection based on fish species has been minimal, which may explain their lack of involvement. When they are involved, however, internationalization becomes more likely.

This latter phenomenon may account for the otherwise puzzling inclusion of Antarctic fishery regulations in the list of internationalized regulations. Although U.S. fishers bear small costs from the mostly forward-looking Antarctic fishery regulations, there was some potential for U.S. fishers to suffer future disadvantages in competition with states that had not agreed to uphold international regulations. Alone, this incentive might not have been a sufficient reason for international efforts to convince more states to undertake obligations, but in conjunction with the importance of the ecosystem as a whole, it gained increasing salience. Moreover, the internationalization of the protection of Antarctic fishery resources followed precisely the model set by earlier efforts (most notably the Tuna Conventions Act and the Pelly Amendment) to ensure that foreign fishers would be bound by the same international regulations that U.S. fishers were held to. In the same way some of the endangered species regulations built on existing internationalization legislation without having to create an entirely new coalition to support it, the Antarctic Marine Living Resources Convention Act internationalized protection of Antarctic fisheries through a broader coalition of actors. It is also worth noting that this legislation, without much of an industry coalition partner, has some of the weakest sanctions provisions in all of the regulations considered here.

On the whole, economic externalities seem to be the most important consideration in determining which fishery regulations the United States works to internationalize. But the entire set of cases is best understood when we consider the advantages that environmentalists add to the attempt to convince the U.S. government to persuade other states to adopt the same types of regulations to which U.S. fishers are already subject.

## Stage I Conclusions

It is not surprising that in the issue areas examined in chapters 3, 4, and 5 the strongest evidence for the primacy of environmental causes of efforts at internationalization comes from the area of endangered species regulations, and the strongest evidence for the primacy of economic causes of efforts at internationalization comes from the area of fishery regulations. When examining the three issues areas together, however, the overall argument that coalitions of environmentalists and industry actors are the driving force behind internationalization is most persuasive. If competitiveness were individually the cause of internationalization, then a greater number of the air pollution regulations should have been internationalized. If environmental considerations were the major factor, then we would have expected a greater number of internationalized regulations pertaining to endangered species. All the regulations that the United States decides to internationalize have their roots in the policy aims of a particular set of environmentalists and industry actors who wanted their internationalization. Issues like protection of elephants or whales, that do not appear to have any competitive advantage incentive for internationalization, are nevertheless internationalized using legislation passed by Baptist/bootlegger coalitions on other issues.

Occasionally, it can be difficult to determine whether environmental or economic factors are those providing the most important incentives, since both are often present. Indeed, they are sometimes represented by the same evidence. For instance, in fishery regulations it can be seen as a character of the transboundary (environmental) aspect of regulation to determine which states do not participate in a regulatory system set up at the international level, but the nonparticipation of these states, if they compete with U.S. fishers economically, can be seen as evidence of an economic incentive for internationalization. It is precisely this overlapping concern, however, that leads to the common interests of environmentalists and industry actors.

A coalition of Baptists and bootleggers is thus a necessary (though perhaps not sufficient) condition to explain internationalization. On the one hand, none of the cases of internationalization examined here can be convincingly explained without reference to both sets of actors. On the other hand, legislation to impose trade restrictions on products made in a way that does not meet U.S. air quality standards was supported by a set of industry actors with environmental justification, and the legislation did not pass. The mere conjunction of interests will not always guarantee that the U.S. policy process will take action. But that conjunction certainly makes action more likely.

For almost any regulation that harms economic competitiveness, industry actors will work for internationalization, if they cannot remove the regulation in the first place. This study does not examine all domestic efforts following the existence of the U.S. regulation, but one should not forget that domestic industries often work for the repeal of a regulation before they try to internationalize it. Nonetheless it is because in internationalization their interests intersect with environmentalists—and because, until 1995 and the Republican revolution in Congress, the removal of environmental legislation in the United States was rare—that this strategy proves the most fruitful. Environmental actors can be found to support a wide variety of causes, no matter how obscure or (relatively) unimportant. Simply noting that there are environmental and economic reasons to internationalize may therefore not be sufficient to predict attempts at internationalization. It is, however, the best predictor discovered so far.

More importantly, the evidence shows that, no matter how pressing the environmental reason to expand domestic legislation, internationalization is unlikely to happen unless there are economic interests to push the internationalization as well. The two efforts of internationalization that made it through the domestic process with the least economic support, the African Elephant Conservation Act and the Antarctic Marine Living Resources Convention Act, contain the weakest provisions of all the legislation considered. Baptists need bootleggers. This observation should provide a lesson to environmentalists about which types of international regulations they are most likely to be able to push successfully. It may also provide a strategy for environmentalists who want to work for international environmental regulation. If they work on the domestic level first and pass rules with competitiveness implications, they create allies for the struggle to apply the rules internationally. Regulate locally, expand globally.

Bootleggers, likewise, may not exist without Baptists. The two groups not only need each other, but mutually create each other. Although the story of internationalization told here focuses on the important role of industry, it is the scientific community and the environmental activists who create the drive for domestic regulation in the first place. Such a drive may even first start on a grassroots level, with consumers who want dolphin-safe tuna or who choose to buy spray cans that do not deplete the ozone layer. The way those demands are addressed, both by industry and by governmental actions, sets in motion a process that can ultimately lead to international environmental regulation. The domestic regulations enacted determine which industry actors will be affected by environmental protection and in what way. Thus, the regulations advocated initially will determine future coalition partners.

The partners in the coalition that results are as wary of each other as the original Prohibition partners were. Both have a number of interests that do not intersect. Baptists may benefit from the economic clout of bootleggers but would prefer to do away with their activity altogether. Bootleggers may gain from the respectability of Baptists but fear their righteous fervor. In the internationalization of domestic environmental policy, this disjuncture became clearest in the case of ozone layer protection. Although there was a brief confluence of interests when both industry and environmental actors wanted to make sure foreign producers and consumers were regulated, in the long run the bootleggers had to close down (or radically change) their business. This type of result may make industry actors wary in the long run of advocating international regulation, particularly when a given industry faces multiple environmental issues. The Alliance for

Responsible CFC Policy has now become the Alliance for Responsible Atmospheric Policy, addressing issues of climate change more broadly. It is likely to have become more wary of cooperation with environmentalists, and of international regulation generally, from its ozone protection experience.

The cases examined here help to isolate the aspects of economic or environmental externalities that are important. On the economic side, there is little evidence for the importance of competitive technological advantage. Even on issues where there was transboundary environmental damage and competitive advantage in regulation technology, such as acid rain, the competitive advantage experienced by manufacturers of scrubbers could not balance the economic harm feared by industries from the stricter regulation that international cooperation would bring. In most environmental issues the regulated industry will be larger than the industry spawned to create technological solutions to environmental problems. In the case of ozone depletion the technological solutions were important incentives for internationalization, but that may be a special case. The same industries that suffered from the regulations domestically on that issue profited from selling substitutes internationally when the regulations were adopted more widely. It is probably only in these types of cases that competitive technological advantage can play a role, but even in these cases it may not be necessary.

On the environmental side, it is more difficult to isolate the most important factor contributing to the Baptist-and-bootlegger coalition. Aspects of harm and the transboundary nature of the problem both seem to play a role. It is clear, however, that harm actually experienced by U.S. actors and transboundary environmental effects felt on U.S. soil are more powerful motivators than those felt by others elsewhere. Also important, however, is the role of concern by environmental groups and the consensus among scientific actors about the seriousness of the problem. Perhaps this factor should not be surprising, given that these actors are the ones who comprise the Baptists in the coalitions. The evidence does suggest, however, that examining the regulations that scientific and environmental groups push for is more important in understanding internationalization of domestic environmental regulations than is an impartial examination of the character of the environmental issue.

One concern worth allaying is the possibility that the United States attempts to internationalize only those regulations that it thinks it will be able to convince others to adopt. In the first place, there is little evidence from the congressional debates on internationalizing legislation (either the measures that passed or the ones that did not) that this factor is explicitly considered. More important, however, is the fact of variation in the success of the United States in persuading other states to adopt those regulations it attempts to internationalize. That variable rate of success does not prove that potential success is not a consideration at Stage I, but it does show either that actors cannot accurately predict success, or that they work for internationalization for reasons other than possible success. The next two chapters address the U.S. attempts at internationalizing the domestic environmental regulations identified here, and examine the effects of the attempted internationalization. Given the mixed record of success, under what conditions should we expect that the United States will actually succeed in convincing others to adopt regulations similar to those it has on the domestic level?

# Stage II
## Success of Internationalization

# 6

## Internationalization Success: Market Power and Threat Credibility

How successfully are domestic environmental regulations internationalized? What determines how successful an attempt to convince foreign states to adopt regulatory policies will be? If domestic policy action is the starting point of one route toward regulation on the international level, we want to learn not only of the attempts by domestic actors to broaden their national regulation, but whether those attempts actually result in international action. There may be domestic pressure to persuade other states to regulate, but it is only when these target states agree to do so that this pressure actually has an international effect. When does policy convergence across states result from one state attempting internationalization, and why?

Internationalization of domestic policies is not a phenomenon unique to international environmental regulation, and consideration of this phenomenon may contribute to an understanding of regulation at the international level more generally. Examination of environmental cases can help to determine to what extent the characteristics of the issues and the states in question influence the ability of states to internationalize their domestic policies. Is there a simple hegemonic ability to impose policy on others? Are those with greater relative amounts of relevant power resources able to prevail in setting international standards? Or are there other factors that contribute to the ability to influence regulation across borders? What will we need to know about the states and issues involved to ascertain whether or not a state will be able to gain international acceptance of a regulation it has already adopted?

The question of success in internationalizing environmental regulations relates to issues that bear on themes fundamental to international relations. Under what conditions will states be able to achieve their objectives through threats? What power resources are relevant for influencing the policies adopted by other states? How do we achieve lasting change in the actions of states? How does unilateral action relate to international cooperation?

As a result of the internationalization of domestic policies of a state that is an environmental leader, other states may be persuaded individually to adopt convergent regulations, they may negotiate multilateral agreements to pursue regulation on the international level, or they may undertake no change in their policies. If target states do take action in response to an internationalization attempt, they may adopt policies that are designed to be acceptable to the state pushing internationalization while changing their actions as little as possible, or they may embark on lasting changes in the way they do business. The change may be undertaken only as long as a threat by the internationalizing state exists or may be integrated into the way the state operates and may continue long after the internationalizing state stops addressing the issue. It would be a useful addition to our understanding of the international policy process to be able to figure out which domestic regulations states are likely to push at the international level, and whether and in what form these regulations will be adopted internationally.

This chapter examines the conditions under which target states might decide to adopt regulations that another state is pushing. The three main hypotheses for explaining these decisions are the legitimacy of the goals and means used, the factors involving the advantages of the regulations to the target state, and the characteristics of the threat by the sending state. The argument put forth here is that, whereas the advantages of the regulation and the legitimacy of the request are not insignificant factors in target state adoption of regulations, the characteristics of a potential threat by the internationalizing state are the most important factors in explaining the sending-state's success in internationalizing its regulations. Specifically, the relative market power the sending state has over the target state in the resource toward which the regulation is targeted is the best predictor of success of internationalization. Although the state may not adopt a regulation out of any particular advantage gained, the costliness of adopting the measure can be weighed against the cost of possible retaliation for not doing so. Also important is the credibility of the threat. This credibility is represented in part by a coalition of "Baptists and bootleggers," present at the first stage of the process, that gains either from the internationalization or from the imposition of the threat.

First, some definitions: for the purpose of this discussion, those states that the United States attempts to convince to adopt environmental regulations will be considered "target states" because they are the targets of U.S. efforts at internationalization; the U.S. or other states attempting to internationalize will be called "sending states." This terminology is consistent with that used by those who try to understand the use of economic instruments to induce policy change in other states.[1] It is also necessary to define success, as it relates to internationalization attempts. Three issues should be considered in arriving at a definition: recognizing success due both to unilateral and to multilateral efforts, determining success in general, and ascertaining the duration of successful internationalization. At the first stage of internationalization, the decision a state makes to push for foreign adoption of its standards can be implemented with a variety of strategies. In particular, as noted in chapter 2, the United States pushes using a two-pronged strategy involving both multilateral diplomacy and unilateral threats. Sometimes one technique is used to the exclusion of the other, but frequently they are used together. When multilateral diplomacy is used, unilateral threats are often held in reserve, sometimes explicitly, in case multilateral negotiations fail to produce the desired outcome. Attempts at internationalization may, in turn, have three possible outcomes: unilateral adoption or multilateral adoption of U.S. standards, or no adoption of U.S. standards, in the cases where internationalization fails. It would not be surprising to find multilateral agreements resulting from cases where multilateral diplomacy is the main internationalization strategy pursued by the United States. We could also imagine cases in which multilateral agreements result from unilateral action, however. States that are targets of unilateral threats may decide it is to their advantage to move the issue into a multilateral arena where one state is less likely to be able to control the negotiating process. Or, once a target state has decided to regulate, it may decide

---

[1] See, for example, Lisa Martin, *Coercive Cooperation: Explaining Multilateral Economic Sanctions* (Princeton: Princeton University Press, 1992); Gary Clyde Hufbauer, Jeffrey J. Schott, and Kimberly Ann Elliott, *Economic Sanctions Reconsidered: History and Current Policy,* 2nd ed. (Washington, D.C.: Institute for Internatioanl Economics, 1990); sending states are also occasionally referred to as "internationalizing states."

that multilateral regulation is a more reasonable strategy for addressing the environmental problem at issue.

The idea that multilateral negotiations will constrain the ability of one state to direct the process shows the difficulty of addressing the determinants of success multilateral contexts. For a simple determination of success, however, it will suffice to ascertain whether the states that are targets of the internationalization attempt have adopted the regulations in question, either unilaterally or as part of a multilateral treaty.

The success of internationalization in general can be determined in several different ways, two of which are employed here. First, an internationalization attempt can be seen as successful if the United States is so satisfied the state in question has adopted a policy (either unilaterally or through a multilateral agreement) sufficiently similar to its own regulation that it stops pushing for internationalization. In other words, internationalization is successful when the United States stops pressuring a target state for further action. The advantage of such a measure is that it allows us to determine when the actor whose actions initiate the process has concluded that internationalization has been sufficient for its purposes.

This measure of success is unsatisfactory in several important ways. As we saw in Stage I, a number of domestic factors influence the decision by the United States to attempt internationalization in the first place. And as we will see in this stage, international factors also enter into the calculation. It would be possible for the United States to abandon its internationalization attempts for reasons that have little to do with whether the target state has adopted the policy in question. For example, domestic actors within the United States pushing for internationalization may change; or legislation authorizing sanctions may expire and not be renewed; or gaining the cooperation of the target state in a different issue area may be important enough to U.S. negotiators for them to agree to overlook inaction on environmental goals. Thus, the United States has been reluctant to impose economic sanctions on China for lack of progress in preventing trade in endangered species, for instance, when other political priorities prevailed. Or again, the United States removed sanctions on Spain for fishing for yellowfin tuna outside the IATTC once that organization ceased regulating catches. The simple abandonment by the U.S. of its efforts at internationalization toward a particular state may not be sufficient evidence to conclude that initial U.S. goals have been met. Conversely, there are cases where U.S. domestic groups force the continuation of an internationalization policy even when the bulk of U.S. objectives have been met. This dynamic can be seen in the internationalization of dolphin protection policies. On the one hand, most states have taken action that protects dolphins in the course of tuna fishing, in response to U.S. actions. On the other hand, several states have refused to meet the specific standards the United States demanded and are therefore seen by official U.S. policy as not accepting internationalization, even though those states generally protect dolphins.

A second measure of success is necessary, therefore, to ensure that the whims of the domestic U.S. policy process do not bias a determination of success. The second measure involves a judgment about the extent to which policy change in the state in question seems consistent both with the original U.S. regulation that is being internationalized, and with the goals explicitly stated by the U.S. negotiators in the attempt at internationalization. Note, however, that in negotiations, states often put forward positions they do not expect to be adopted, so that a more restrained position, closer to the original goals, can be taken in a compromise.[2] The fit between these two views of the internationalization goal and the policy adopted by the states in question can be analyzed. In practice it is usually not very difficult to determine whether a state has adopted the policy in question.

More difficult may be to determine whether the target states actually follow through with the commitments they make to adopt environmental regulations, either domestically or multilaterally. Certainly there is less meaning to the adoption of a regulation if the state does not then adjust its behavior to the requirements of the new rule. Thorough examination of this level of compliance by all states is impossible for a project of this scope, but the cases where behavioral change is clear or unlikely are noted. Occasionally the internationalization process itself produces such evidence. When the domestic actors that push for internationalization in the first place are not convinced that the target state is complying with the policy it has adopted, their concern can be brought out through the same channels through which internationalization was initially pursued. Additionally, if the internationalization

---

[2]  See, for example, Thomas C. Schelling, *The Strategy of Conflict* (Cambridge, Mass.: Harvard University Press, 1960).

is multilateral, the organization created or used may have some mechanism for detecting or punishing noncompliance.

In a similar vein, we also need to look at both short-term and long-term success. Short-term success involves the target states taking the requisite action at the time they are pressured to adopt regulations. This type of success may be ephemeral, however, if the change is not institutionalized. States that take action only to respond to external pressure may undo that action as soon as the external pressure disappears, particularly if they have not been convinced of the advantage of the regulation they have been pressured to adopt. This problem is not unique to situations in which states face negative pressure for action; the same situation can arise in circumstances in which states are offered incentives to take certain actions. They may cease to take the desired action once the incentive to do so is removed.[3] Long-term success, therefore, involves some policy change by the target state that is unlikely to be undone. Institutionalization in a multilateral agreement might make a state less likely to undo its commitment to the regulation, for instance, as would a change in infrastructure for a production process. Once industry has changed to a new way of doing business, it may be less expensive to continue even after the incentive that created that change has disappeared. For example, states that receive aid under the Montreal Protocol Multilateral Fund to change to manufacturing processes that do not use ozone-depleting substances have little reason to go back to ODS use even without threats or promises since their current equipment uses a different process. Long-term success might also be more likely if the state undertakes the change because it has been convinced that the change may be in its interests, apart from the pressure exerted.

It is worth mentioning another important element of success: the extent to which the environmental situation improves because of U.S. efforts to internationalize domestic environmental standards. It is important not to lose sight of this aspect of success, but it is also difficult to make use of it here. In the first place, the time frame of many environmental issues is such that changes in behavior may not have a measurable effect until twenty or

[3] See, for instance, Paul Mosley, Jane Harrigan, and John Toye, *Aid and Power: The World Bank and Policy-Based Lending*, vol. 1 (London: Routledge, 1991), pp. 30–31; Alex Duncan, "Aid Effectivenss in Raising Adaptive Capacity in Low-Income Countries," in *Development Strategies Reconsidered*, ed. John P. Lewis (New Brunswick, N.J.: Transaction Books, 1985).

thirty years afterward. Ozone-depleting substances can remain in the stratosphere for up to a century, and fishery stocks can take decades to recover. Since policies relating to many of these issues have only recently been internationalized, it is too soon to determine all the environmental effects of internationalization. Second, environmental problems frequently have multiple causes. If sea turtles are less threatened because of the internationalization of shrimp fishing regulations, they may still be harmed by habitat degradation or by use in traditional medical practices.[4] Protection from one source of threat may provide improvement but not solve an entire problem; it is not clear that it would be fair to expect any one policy to accomplish everything.

The definition of success used here is primarily a political one, in part because the process that creates it is political. Those who work for internationalization of a U.S. environmental law have a variety of reasons for doing so, and not all of those reasons relate to the improvement of the environment. This analysis, therefore, takes the definition of success—adoption of a regulation—used by the U.S. actors pursuing it. But it is nevertheless valuable to determine the extent to which the political jockeying examined here is likely to result in any improvement of the environmental problem that, at least in part, motivates it. Otherwise, success in the long run would be meaningless.

The cases considered here are those identified in chapters 3 through 5; instances in the three issue areas examined—endangered species, clean air, and ocean fisheries—in which attempts are made by the United States to convince other states to adopt regulations the United States already has. The success of these attempts is analyzed with respect to the argument developed below.

## Internationalization Success — The Ozone Example

There are a variety of reasons for which a target state might want to adopt the regulations it is pressured to adopt. First, a state may adopt a regulation because it thinks that doing so is the "right" thing, regardless of coercion or benefit. Second, a state might adopt a regulation because the

---

[4] "Turtles: Trade in SE Asia Raises Threat of Extinction," *Greenwire,* 4 May 1999.

regulation itself confers some benefit on the state, particularly if the internationalizing state helps avoid collective action problems in efforts to adopt international environmental policy. Third, a state might adopt a regulation because it fears that it will be harmed if it does not do so. As at Stage I, the wide variety of explanations given for the successful internationalization of regulations on ozone-depleting substances can be illustrative for thinking about success of internationalization overall. Although it would be a mistake to consider the entirety of the negotiations about preventing stratospheric ozone depletion as internationalization of U.S. domestic environmental policies, the United States was a leader in the negotiations, and some of the United States' negotiating position stemmed from a desire to involve other states in domestic actions it had already taken or was considering. If our starting point is domestic regulation, an examination of domestic U.S. regulations on ozone-depleting substances followed forward shows pressure for internationalization of these standards. This chapter therefore continues to consider the ozone negotiations as an attempt at internationalization of a domestic policy, and uses them to illustrate possible reasons for success of internationalization. The actual explanations for success examined in this chapter, however, have been derived from international relations theory more broadly.

A factor considered important by some as an explanation for success of the ozone negotiations is the legitimacy of the process by which the United States worked to internationalize its domestic regulation. Because the United Nations Environment Programme coordinated the negotiations, the process was seen as having international support and additional credibility.[5] That the Multilateral Fund was created to give positive encouragements rather than negative sanctions for joining the agreement was also seen as an internationally legitimate way to go about broadening international participation in regulation of ozone-depleting substances.[6]

The most prevalent explanation for the successful internationalization of regulations preventing ozone depletion has to do with factors relating

---

[5] Richard Elliot Benedick, *Ozone Diplomacy: New Direction in Safeguarding the Planet* (Cambridge, Mass.: Harvard University Press, 1991), p. 5.

[6] Diane Dumanski, "In Shift, U.S. to Aid World Fund on Ozone," *Boston Globe*, 16 June 1990: 1; Friends of the Earth, *Funding Change: Developing Countries and the Montreal Protocol* (London: Friends of the Earth, 1990); Benedick, *Ozone Diplomacy*, p. 152.

to the environmental damage of ozone depletion: the necessity to prevent global damage through international action. The increasing scientific evidence that ozone depletion causes environmental damage contributed to the call for international action. The discovery in 1987 of a thinning of the ozone layer over the Antarctic is often seen as the factor that led to the signing of the Montreal Protocol.[7] The increasing level of information about the seriousness of the damage to the ozone layer clearly influenced the willingness of the international community to adopt regulations to prevent it.[8] U.S. research that contributed to the domestic regulation in the first place continued as a condition of the domestic regulation,[9] and, along with international research programs, helped to gather greater evidence of the seriousness of the problem. The U.S. Environmental Protection Agency did a study of the risks and effects of ozone depletion, which European states used in considering ozone policy.[10]

In addition, the United States acted as a coordinator of collective action, which allowed individual states to reach the collectively preferable outcome of joint regulation it might have been difficult for states to approach individually. Further, U.S. regulations also had a demonstration effect, indicating that substitutes for ozone-depleting substances could be found and used.

[7] See Peter Morrisette, "The Evolution of Policy Responses to Stratospheric Ozone Depletion," *Natural Resources Journal* 29 (1989): 814–5; Steven J. Shimberg, "Stratospheric Ozone and Climate Protection: Domestic Legislation and the International Process," *Environmental Law* 21 (summer) (1991): 2188–89; Karen T. Litfin, *Ozone Discourses: Science and Politics in Global Environmental Cooperation* (New York: Columbia University Press, 1994), pp. 79–80. It is not actually clear what role the "ozone hole" played in the negotiations; the announcement of the discovery was not made until after the Montreal Protocol had been signed, and knowledge of the finding came to the negotiators late in the process. In addition, when the "hole" was discovered it was not immediately clear that it was due to anthropogenic factors. Benedick, *Ozone Diplomacy,* p. 108. In contrast, Litfin argues that the negotiators did know about the "hole" and considered it in their negotiations even when they had decided not to.

[8] Morrisette, "Evolution of Policy Responses," pp. 812–25; Benedick, *Ozone Diplomacy,* p. 5.

[9] See Robert Stewart, "Stratospheric Ozone Protection: Changes over Two Decades of Regulation," *Natural Resources and the Environment* 7 (2) (1992): 24–27, 53–54; Orval Nangle, "Stratospheric Ozone: United States Regulation of Chlorofluorocarbons," *Environmental Affairs* 16 (summer) (1989): 531–80; Morrisette, "Evolution of Policy Responses," p. 813.

[10] Morrisette, "Evolution of Policy Responses," p. 813.

Moreover, that U.S. actors already had to do without ozone-depleting chemicals in some applications led to the creation of substitutes that made international adoption of regulations more feasible.

A final set of explanations for the international adoption of regulations on ozone depletion focuses on the relative power of the various actors involved and the threat that the power will be used. The important world position of the United States contributed to its ability to convince other states to join in international negotiations to address ozone depletion.[11] The dominant market position of the United States in ozone-depleting substances meant that if it instituted sanctions in the event that multilateral negotiations failed, these sanctions could impose some harm on the target states, and it made that possibility explicit through discussion of legislation pending in Congress. The trade sanctions in controlled substances adopted under the Montreal Protocol gave those who needed access to these substances incentives to join the agreement. The credible threat by the developing states to stay out of the agreement contributed to their ability to gain a compensation package after the negotiation of the original agreement.[12] Power, in other words, was present in the negotiations at Montreal.

To what extent do any of these types of stories about international negotiation of ozone layer protection explain the general success of internationalization? What is known about the conditions under which one state will be able to persuade others to adopt a regulation it has already adopted?

### Legitimacy

A factor that has been seen to play a role in the successful persuasion of states to adopt policies they did not originally have is the legitimacy of the request. Thomas Franck's investigation of the "power of legitimacy" begins by seeing it as "a property of a rule or rule-making institution which itself exerts a pull towards compliance on those addressed normatively."[13] To what

---

[11] Benedick, *Ozone Diplomacy*, pp. 55–58; Benedick also points out the potential Reagan administration reticence late in the negotiations threatened to undermine them, given the important role of the United States, p. 58 ff.

[12] Elizabeth R. DeSombre and Joanne Kauffman, "Montreal Protocol Multilateral Fund: Partial Success Story," in *Institutions for Environmental Aid: Pitfalls and Promise,* ed. Robert O. Keohane and Marc Levy (Cambridge, Mass.: MIT Press, 1996).

[13] Thomas M. Franck, *The Power of Legitimacy among Nations* (New York: Oxford University Press, 1990), p. 16.

extent is the initiating state seen to have a right to make the demand on the target states? To what extent does it follow the procedures of international law in working to convince other states to adopt similar regulations? Elements of legitimacy thus can come either in the request itself or in the process followed in making that request. It can be important even for powerful states to act in legitimate ways. Abram and Antonia Chayes argue that U.S. military or economic power "might be undermined if it did not seek and achieve a degree of international consensus to give its actions legitimacy."[14]

Those who argue that international law or even the workings of international society have an effect on the actions of states believe that states are more likely to respond to internationally legitimate requests. One aspect of this issue is the legitimacy of the issue considered for internationalization. This legitimacy can be represented by the extent to which the international community has agreed on the goals of regulating internationally on the issue in question. Likewise, the existence of a prior international agreement that the internationalizing state is hoping to expand or strengthen could give evidence of the acceptance by the international community of the legitimacy of an issue and thereby provide additional pressure on the target of internationalization.[15]

Also at issue is the extent to which the method of persuasion is considered to be legal or legitimate. The difference between pursuing internationalization via multilateral or unilateral measures could be important if legitimacy plays a role. If unilateral measures are threatened, the extent to which those measures are consistent with other regulations in the international arena can make a difference. One of the most frequently voiced concerns over the United States' threat and use of trade sanctions under Section 301 (legislation authorizing unilateral economic action similar to that examined in the environment cases, used for pursuing trade liberalization)

[14] Abram Chayes and Antonia Handler Chayes, *The New Sovereignty: Compliance with International Regulatory Agreements* (Cambridge, Mass.: Harvard University Press, 1995), p. 41.

[15] See, for example, Elisabeth Zoller, *Enforcing International Law through U.S. Legislation* (Dobbs Ferry, N.Y.: Transnational Publishers, 1985), especially p. 84ff. She refers to the role of unilateral activities as used to "exercise enough pressure to prevent a foreign state from jeopardizing specific collective interests," p. 93. See also Chayes and Chayes, *New Sovereignty,* p. 129.

is that U.S. actions sometimes violate international law. In some of these instances GATT dispute-settlement panels have pronounced U.S. actions illegal. Jagdish Bhagwati, among others, argues that using illegitimate methods of persuasion are likely to be counterproductive in these cases, because states will resist pressure they consider to be illegal and may even retaliate.[16] Additionally, if international law plays a strong role in internationalization and international dispute resolution procedures have ruled that a measure is not legal, it should be abandoned, thus derailing the attempt at internationalization. There is therefore reason to suspect that internationalization attempted through instruments seen as illegitimate will be less successful than those relying on legitimate ones, and that internationalization attempted by measures ruled illegal by international bodies should be the least successful of all.

An explanation of the success of internationalization based on legitimacy would predict the greatest success when the goals, and the means by which those goals are pursued, are seen as legitimate by the target state and the international community. As Franck points out, it is difficult to isolate a belief in legitimacy from other motives.[17] Nevertheless, it might be possible to infer evidence for the importance of legitimacy from the extent to which the target states or the rest of the international community question the acceptability of internationalization goals or the means by which they are pursued. Also important would be the extent to which the target states pursue international relief for the pressure inflicted on them. If legitimacy matters and U.S. actions are ruled illegal, we would expect those illegal internationalization actions to be less successful than others seen as more legitimate. Yet the United States has refused to drop its push for internationalization of regulations when the means have been ruled illegal, showing quickly that legitimacy is not likely to be a primary factor in the success of internationalization. Alan Hyde has indicated that every study he located "that attempted to show a relationship between an attitude or belief in the legitimacy of an order and some corresponding behavior has found that the relationship is weak or

---

[16] Jagdish Bhagwati, "Aggressive Unilateralism: An Overview," in *Aggressive Unilateralism: America's 301 Trade Policy and the World Trading System*, ed. Jagdish Bhagwati and Hugh T. Patrick (Ann Arbor: University of Michigan Press, 1990), pp. 33–35. Under some rules, including that of the GATT, the aggrieved party may be legally allowed to retaliate.

[17] Franck, *Power of Legitimacy*, p. 19.

nonexistent."[18] That means ruled illegal have nevertheless induced change in target state behavior supports Hyde's observation.

## Self-Interest

An important reason target states might accept internationalized regulations is the benefit these regulations confer. Certainly the damaging effects of international environmental degradation provide an incentive for states to regulate, particularly collectively, to prevent environmental harm. One set of explanations for the success one state has in internationalizing regulations therefore focuses on the advantages for the target state of adopting the regulations in question: for instance, states adopt similar standards for technology to reduce transaction costs in international trade; they determine standard languages and procedures for air traffic control to facilitate international travel; they regulate open-ocean fishing to ensure the indefinite continuation of fishing. Because of the voluntary nature of international law, states adopt regulations when they have decided it is to their advantage to do so.[19]

Although at the first stage of this process environmental factors alone were not shown to be the determining influence on decisions to internationalize, we should not necessarily bar them from our consideration at this stage. The two stages involve a different set of decision makers. At Stage I U.S. actors consider the environmental advantage to the United States. At Stage II actors in the target state decide about whether to accept U.S. pressure and to adopt a given regulation; for them the environmental costs and benefits may be entirely different than for the U.S. actors.

The idea that states adopt environmental regulations because it benefits them environmentally to do so is such a basic assumption that empirical investigation often simply takes it for granted. The "concern" that Peter Haas, Robert Keohane, and Marc Levy see as a necessary ingredient for creating international environmental institutions is concern about environmental damage.[20]

---

[18] Alan Hyde, "The Concept of Legitimacy in the Sociology of Law," *Wisconsin Law Review* (March/April 1983): 395.

[19] No state can be legally compelled to enter into an international agreement, save possibly a peace treaty. See Michael Akehurst, *A Modern Introduction to International Law,* 6th ed. (London: Harper Collins Academic, 1987), pp. 123–42.

[20] They note that the first stage in regulation is "agenda setting," which begins with environmental problems that are recognized as needing action. Robert O. Keohane, Peter M. Haas, and Marc A. Levy, "The Effectiveness of International Environmental Institutions," in Haas, Keohane, and Levy, *Institutions for the Earth,* p. 12.

Investigations of international regulation do not assume that environmental damage will inevitably lead to international action to mitigate it, but they seldom find regulation without environmental harm.[21]

A first effort to construct an environmental explanation of the success of internationalization may have more to do with environmental conditions than with pressure per se. The extent to which the state in question suffers from an environmental problem may influence its likelihood of implementing internationalized regulations.

The simplest version of an explanation attributing successful internationalization to environmental effects would predict that the extent to which a target state is harmed by the environmental damage in an issue area would determine its likelihood of adopting regulations pushed by the United States. In other words, not all states are equally affected by all environmental problems; those that bear the greatest environmental consequences from a given environmental issue should be the most likely to adopt regulations to address these issues. There may be some difficulties in applying this observation across issues as well as across states. It may be simple to determine that Ecuador suffers greater loss from the depletion of tuna fisheries than does Venezuela, or even that Ecuador suffers more from tuna overfishing than it does from air pollution, but it is much harder to assess whether Ecuador's air pollution is worse than Venezuela's fisheries problem. Nevertheless, we would expect to see more target state acceptance of internationalization in instances where the environmental harm to that state is high.

In the set of cases considered here, however, self-interest of the target state is unlikely to be the only explanation for success of internationalization efforts. If regulations were a pure gain for the target states, the states would probably already have adopted them. This logic is similar to that employed by economists who question the "economic gain from environmental regulations" argument described in chapter 2. If a regulation is a pure positive gain, then rational actors would be able to figure that out and would already have adopted it. Explaining why actors had not previously adopted a policy that would be advantageous requires focusing on the

---

[21] Andrew Hurrell and Benedict Kingsbury, "The International Politics of the Environment: An Introduction," in *The International Politics of the Environment*, ed. Andrew Hurrell and Benedict Kingsbury (Oxford: Clarendon Press, 1992), p. 47.

factors that prohibited the actors from realizing the advantages of the regulation originally. On the international level there are even more such factors than there are likely to be on the firm level.

Policies that are advantageous to states individually or collectively may nevertheless be difficult to adopt in international regulation in general, and more difficult in regulation pertaining to environmental protection in particular. These difficulties may be structural or social. Structurally, actors wishing to cooperate in an anarchic system have no way to guarantee that cooperation will succeed, and for that reason alone it may fail. In Hardin's "tragedy of the commons," cow herders would all be better off if they were able to take regulatory measures to prevent the depletion of the commons; they are unable to do so because the lack of enforcement gives advantages to those who free ride and the incentive to everyone to do so, thereby ruining the commons.[22] The prisoner's dilemma provides a similar analogy. Despite the fact that cooperation would produce an outcome better for both, two prisoners are unable to make a successful deal because neither has assurance that the other will not defect. To protect themselves from that possibility, both defect, and end up with an outcome that is worse for both than if they had cooperated. In some cases, if states are concerned about their relative positions, cooperation is even less likely, even if all states involved would benefit absolutely from the interaction. For structural realists this concern with relative gains is the main reason that meaningful international cooperation is unlikely.[23] The security dilemma and resulting arms races during the cold war are one example of this phenomenon in practice on the international level.[24] The lack of universal free trade is another.[25]

More generally, problems of collective action on the international level are blamed for the difficulty in gaining international action to address

[22] Garrett Hardin, "The Tragedy of the Commons," *Science* 162 (13 December 1968): 1244.

[23] See, for example, Kenneth Waltz, *Theory of International Politics* (New York: Random House, 1979), pp. 105–7.

[24] See Robert Jervis, *The Illogic of American Nuclear Strategy* (Ithaca: Cornell University Press, 1984).

[25] Waltz, *Theory of International Politics,* p. 104; John Conybeare, "Trade Wars; A Comparative Study of Anglo-Hanse, Franco-Italian, and Hawley-Smoot Conflicts," in *Cooperation under Anarchy,* ed. Kenneth Oye (Princeton: Princeton University Press, 1986), p. 152.

problems cooperatively. One such problem is that of free riders. Although all states might benefit from a collective solution that involves small sacrifices for great joint gains, each benefits more individually if all other states make the sacrifices while it does not. In the individual calculation, the free rider shares in the collective benefit from the restraint of the rest of the community while bearing none of the costs of changing behavior. But because in this formulation all actors involved have similar incentive structures, when each one tries to free ride there is no cooperative outcome.[26]

In addition to the international problems considered above, structural problems on the state level may hamper the ability of a state to adopt regulatory policies that might benefit it or some of its constituent parts. Although states are often considered unitary actors within international politics, the positions they take, as illustrated at Stage I, arise from preferences of actors within their borders. The way those preferences are aggregated may give priority to certain types of or conjunctions of interests. Environmental externalities produced by industrial practices often harm large but dispersed populations that are unlikely to be able to organize successfully to call for a change. A small, well-organized set of actors gain the advantages of these industrial practices. Even if the benefits to society from regulation would outweigh the costs borne by industry, the smaller number of individuals harmed, and the greater degree of harm per capita, gives industry superior organizational advantages to prevent regulation from being imposed.[27]

Another structural factor preventing a state from adopting a beneficial regulation might be the time horizons of the various actors. States—and other actors—may not take action that might benefit them in the long run because of the short-term advantages of not acting. The emission of greenhouse gasses is one example of a situation in which the harm from unregulated activities will arrive sufficiently far in the future so that current politicians may

---

[26] See Mancur Olson, *The Logic of Collective Action: Public Goods and the Theory of Groups* (Cambridge, Mass.: Harvard University Press, 1965); Robert Axelrod, *The Evolution of Cooperation* (New York: Basic Books, 1984). It is for this reason that many assume that a hegemon is necessary to overcome these types of collective action problems. See, for example, Robert Gilpin, *U.S. Power and the Multinational Corporation* (New York: Basic Books, 1975), p. 85.

[27] This logic is similar to that derived by Olson, who argues that "members of 'small' groups have disproportionate organizational power for collective action." Mancur Olson, *The Rise and Decline of Nations: Economic Growth, Stagflation, and Social Rigidities* (New Haven: Yale University Press, 1982), p. 41.

be unlikely to want to address it. The political system may play a factor in this sort of decision. Politicians running for election are likely to be concerned with policies that provide benefits in the short term; actions that might be of greater advantage overall but less obviously advantageous immediately are less likely to be popular. Some of this tendency can be attributed to the whims of political processes and electorates, but discounting the future may be a rational response to uncertainty. Trading uncertain future harm for certain current benefits is a common political choice.

Uncertainty more generally can be seen as another contributing factor in preventing a state from adopting a policy that might be advantageous. Often environmental damage is surrounded by uncertainty: To what can the damage be attributed? How serious is the environmental harm? What are the possibilities for addressing it? To what extent do scientists understand and agree on the danger from the environmental problem? Uncertainty has certainly played a major role in environmental regulation on both the local and the international levels.[28]

The structural factors above address the way a given set of preferences is arrayed. On a social level, the preferences of the actors in question might actually change. A set of actors previously not in favor of environmental regulation might decide to reassess its position. This phenomenon is sometimes considered "learning"; it can also encompass the social construction of norms and preferences.[29] It can be difficult to draw a clear line between structural and social factors. If resolving uncertainty about the causes or seriousness of a problem leads actors to want environmental protection when they did not previously, have their preferences changed or has the resolved uncertainty simply allowed them to take an action they preferred all along? Whatever the cause, however, actors may have a preference for the environmental regulation that has not yet been realized through the existing political process.

It is therefore possible that a variety of regulations could be beneficial to a state as a whole or to large portions of its population and nevertheless

[28] See Steinar Andresen, "Science and Politics in the International Management of Whales," *Marine Policy* 13 (April) (1989): 99–117; Walter A. Rosenbaum, *Environmental Politics and Policy,* 3rd ed. (Washington, D.C.: CQ Press, 1995) pp. 75–76; 161–90.

[29] See, for example, Audie Klotz, "Norms Reconstituting Interests," *International Organization* 49 (summer 1995): 451–78.

not be undertaken. The simple fact that these regulations have not yet been adopted does not mean that they would not have advantages. In the cases examined here, one state has already begun the process of regulation, which can contribute to the likelihood of other states adopting such regulations by addressing the problems of collective action considered above. If the environmental regulations provide some benefit to the state being pressured to adopt them, one important state's already having such regulations can help overcome potential impediments to regulation.

In terms of overcoming problems of collective action, the presence of regulation by a major actor can provide structural assistance in regulation. Some international relations theorists believe that international public goods or international cooperation can best (or only) be achieved with the assistance of a hegemonic actor.[30] If the action taken by the initial regulator is sufficient to begin to address the environmental problem, its initial action is not useless environmentally, even if international regulation would be required to address the problem fully. Most importantly, though, the act of being the first regulator then gives the initiating state an incentive to act to help other states overcome the problems of collective action that may prevent them from regulating. Because the initial regulator already has a stake in the system, by having sacrificed in order to regulate, it then gains an incentive to provide enforcement in a system where lack of enforcement might prevent cooperation even among states that wanted it. Regulation by another state can also help overcome problems of domestic collective action by playing a coordination role. The action of the first regulator identifies a specific mode of regulation (among many) and thus prevents domestic conflicts about how to regulate.[31]

The presence of pressure from an initial regulator can also work to shift incentives, or the coalitions of actors, within the target state. If the way that

[30] See Robert O. Keohane, *After Hegemony: Cooperation and Discord in the World Political Economy* (Princeton: Princeton University Press, 1984), p. 31, 135ff.; Robert Gilpin, *The Political Economy of International Relations* (Princeton: Princeton University Press, 1987), p. 119; Stephen Krasner, "State Power and the Structure of International Trade," *World Politics* 28, no. 3 (1976): 317–43.

[31] This situation can arrive in games of coordination. See Duncan Snidal, "Coordination versus Prisoner's Dilemma: Implications for International Cooperation and Regimes," *American Political Science Review* 79 (1985): 923–42; and Arthur A. Stein, "Coordination and Collaboration: Regimes in an Anarchic World," *International Organization* 36 (1982): 299–324.

preferences are aggregated gives greater influence to those who gain from the environmental damage (as explained above), the outside pressure that targets those who thus gain may not so much change their preferences as change the opportunities for domestic coalitions within the target states. Public opinion and transnational action by nongovernmental actors may provide another incentive for states to adopt the regulations they are being pressured to adopt; one important state already having undertaken regulation on a given issue may contribute to international public opinion in favor of regulation. Many studies of environmental regulation attribute action to public opinion;[32] it can provide pressure from the inside (while at the same time the state is receiving pressure from the outside).

Uncertainty can also be addressed through action by an initial regulator. If we assume that the damage an environmental problem causes affects a country's decision about whether to adopt an environmental policy, then understanding the effects of a problem can be of central importance. The regulations already adopted by an influential state can address the issue of uncertainty in several ways. In the first place, the initial regulator can provide evidence of the seriousness of the environmental problem. To some extent the fact that one state was willing to pass regulations unilaterally shows that it considers the problem to be significant, especially given that these regulations generally impose a cost on domestic actors. More importantly, environmental regulations either on the domestic or international level often bring with them increased research—sometimes even as part of the regulatory authority itself. The initial regulator may thus be able to produce evidence of the seriousness of the environmental issue. Another uncertain factor in environmental regulation concerns the outcomes of regulation. There can be fear that the regulations will not actually address the environmental problem, or that the regulations will be too costly to implement. If costly adjustments are required to manage environmental problems, it is important to know that making these adjustments will actually lead to

---

[32] See, for example, Francis Cairncross, *Costing the Earth: The Challenge to Governments, the Opportunities for Business* (Boston: Harvard Business School Press, 1991), pp. 2–6; Lynton Keith Caldwell, *International Environmental Policy: Emergence and Dimensions* (Durham, N.C.: Duke University Press, 1984), p. 12. Public opinion can also be seen to be a part of "concern," one of the three C's seen as necessary for international environmental action in Haas, Keohane, and Levy, *Institutions for the Earth*.

environmental improvement.[33] The initial regulator can demonstrate by example the ways to regulate, including new technological processes, and can demonstrate the effectiveness of the regulations. The initial regulator's addressing uncertainty can therefore play a role in the willingness of target states to adopt the environmental regulations in question. An explanation of successful internationalization based on lowering uncertainty is more likely to account for differential rates of success between issue areas than for differential rates of success across target states within one issue area, since the overall level of understanding about an issue area is unlikely to differ across states.

A related approach would focus less on the intrinsic benefits of the regulation than on the costs of adopting it. Although economists would see both these aspects as the same measure (net benefit), the two may represent different decision-making processes, depending on where the costs and benefits fall. A measurement of cost would take into consideration the price of any required technology to meet the standard, as well as any lost income resulting from, for instance, the necessity of using a less efficient production process. Regardless of whether a state sees an intrinsic advantage from the regulation, it would not be surprising if the state were less resistant to regulations with lower costs than to those that would be more expensive to implement.

These explanations based on self-interest express the idea that "success" in internationalization may not be the realization of a sending state's pressure, but may instead represent the decision by target states to adopt regulations on the basis of the intrinsic costs and benefits of those regulations. In the truest form of an explanation based on self-interest, states take actions that benefit them environmentally, but at the very least, the trade-off between the costs and benefits of regulation is likely to be important. At base, explanations for success that involve either environmental effects or costs of regulations assume that the target state adopts regulations because they advantage that state in some way, rather than adopting the regulations only because of pressure. Both elements may in fact play a role in adoption, as the discussion of threat will demonstrate. But while environmental

---

[33] See Marc Levy, "Political Science and the Question of Effectiveness of International Environmental and Resource Agreement: A Status Report," Prepared for the Workshop on International Environmental and Resource Agreements, Fritdjof Nansen Institute, 19–20 October 1992. Cited with permission.

factors such as the seriousness of the problem and the advantages of solving problems of collective action are expected to have some influence on decisions of target states, these are not likely to be the primary determinants of success in internationalization attempts, for several reasons.

First, within the issue areas examined here, the sets of cases on which the United States pushes for internationalization are not those that are likely to be of environmental importance to the target states. Although problems of collective action can make beneficial international regulation difficult and may therefore explain the lack of regulation in areas that would benefit states, there is nevertheless something to be said for the argument that if the target states found regulations sufficiently beneficial, they would have adopted them already. An examination of the issues considered here confirms the lack of centrality of the regulations to the concerns of target states. Protection of endangered species is unlikely to be of highest concern, even among other environmental issues, to states that have not reached a relatively high level of development. The value of preserving particular species or species diversity in general can be shown, but it is unlikely to be the greatest among environmental priorities of target states. Air pollution may be serious in the target state, but it is often is a relatively localized phenomenon, so United States–style regulations may not benefit the target state as much as its own way of regulating would. Fishery conservation is the most likely among the issues examined to benefit the target states, and states have approached fishery conservation by a variety of means not limited to international regulation. It might seem that the lack of interest of target states can be attributed to the case selection, but many other sets of regulations that could have been selected—water pollution, pesticide regulation, waste disposal—would not be of any greater concern to target countries. This observation ties into a larger issue of how environmental issues are taken up internationally: those regulated internationally tend to be those that benefit the most powerful countries, either environmentally or economically.[34]

It is not a coincidence, however, that the cases in which the United States is working to gain regulation by other states are not environmentally significant to the targets. The cases considered here begin on the U.S. domestic level so they are issues about which the United States was strongly concerned.

[34] See, for instance, Gregg Easterbrook, "Forget PCBs. Radon. Alar." *New York Times Magazine,* 11 September 1994: 60–63.

There is no inherent reason that target states are going to be harmed by the same environmental problems that caused the United States to take domestic action.

More importantly, though, given how the decision is made at Stage I to push for internationalization of the regulation, there is even less of a link to environmental damage in the target states than we would predict from a random sample of environmental regulations. Although the decision to regulate initially at the domestic level is made for largely environmental reasons, the decision to internationalize a regulation comes in large part because of the economic advantage domestic industry gains by it.[35] Even if aspects of air pollution, species preservation, and fishery conservation were likely to be important environmental considerations for the target states, the specific regulations within these issue areas that are candidates for internationalization are determined by U.S. industry interests and therefore are even less likely to be of central environmental concern in the target states. It should not, therefore, be surprising if environmental factors do not determine the success of internationalization.

The inability to attribute success in internationalization directly to environmental factors does not mean that environmental self-interest is never important in international environmental regulation. Indeed, it is likely to be important when states decide, without pressure, to work for international regulation. Further, it certainly plays a role, even when there is pressure, in situations where it is likely that states will continue environmental regulations once outside pressure has disappeared. Thus, among the sets of regulations that begin on a domestic level and are candidates for internationalization, the relative degree of concern about the issue (or environmental benefit from the regulation) may still play some role in a state's willingness to accept international pressure.

The cost to the target state of adopting a regulation will certainly play into its decision about whether to accept internationalization of a U.S. policy, but cost in the absence of clear environmental benefit is likely to play a less important role. When choosing a course of action, most actors will do some sort of cost-benefit analysis, but they first need to be persuaded to choose a course of action, and cost alone does not appear to determine that decision.

---

[35] This advantage comes from either the internationalization itself or the economic measures imposed if the target state fails to accept the regulation.

## Threats

A final set of explanations for the propensity of target states to adopt regulations that other states push focuses more directly on the pressure exerted by the internationalizing state. At the margin, these explanations could also be considered explanations based on self-interest. In response to a threat a state may decide that it is better off adopting the regulation than experiencing the consequences of action by the sending state. The explanations based on threat considered here, however, differ from those based on self-interest discussed above. When threats are the primary motive for a state to adopt environmental policies, the target state may not actually gain in an inherent way from the policy. That lack, in turn, may have implications for the likelihood that the state will keep the policy when the pressure disappears. If a state finds that a policy it was pressured to undertake nevertheless provides benefits, it will be more likely to continue it when there is no pressure. There may be cases, however, in which a state that did not find the change to be particularly advantageous would continue it after the pressure is removed, for example, if the policy ushered in a new production process that, once in place, was costly or difficult to change. This comparison points out the importance of seeing self-interest as a continuum: since states will adopt only policies they choose to, their choices may be due to a combination of inherent self-interest and the self-interest created by outside pressure.

Explanations for internationalization based on threats generally involve either the power of the states involved or the costs of implementing the threat in question. We should not overlook the role that power plays in international regulation. Many see it as the main determinant of action on the international level.[36] It is not hard to imagine why a more powerful state might be able to convince a weaker state to adopt particular policies. Abram and Antonia Chayes conclude in their recent compliance study that the imposition of sanctions is largely "hegemons—and particularly the United States—enforcing hegemony."[37] There is no question but that hege-

---

[36] See, for example, Waltz, *Theory of International Politics;* Hans J. Morgenthau, *Politics among Nations: The Struggle for Power and Peace,* 4th ed. (New York: Alfred A. Knopf, 1967); Robert Gilpin, *War and Change in World Politics* (Cambridge: Cambridge University Press, 1981).

[37] Abram Chayes and Antonia Handler Chayes, "Extra-Treaty Sanctions," Draft, December 1, 1992, as presented at the Harvard/MIT Seminar on International Institutions and Political Economy, 10 December 1992. Cited with permission.

mony plays a role in the imposition and success of the sanctions considered here. However we define power, the United States is more powerful than the states it threatens in this issue area.

Power is a complex and multifaceted concept, and it is important to be specific about the different factors that might be relevant in considering power as an explanation for anything. Some see power as "the ability to achieve one's purposes or goals," focusing on the results of power.[38] But such a definition only tells us post hoc whether a state has the necessary power resources (when we discover whether or not it accomplished its goals). Moreover, this approach does not explain which aspects of power will be important to predict which states will prevail in which situations. It would be useful to discover what types of power would be necessary or sufficient for a particular state to achieve a particular type of internationalization goal.

Others focus on resources and attempt to delineate a set of resources that make a state powerful, such as military strength, population, territory, natural resources, economic size and growth, political stability, and even culture, ideology, and institutions.[39] A concept of what constitutes power can help in making judgments a priori about which states are likely to prevail in situations where internationalization is pressed, although the varied and changing definitions of what might be considered important in determining a state's "power" can be unwieldy. When examining the success of U. S. policies, however, some argue that it is not necessary to further refine the definition of power. Particularly in this set of issue areas, the threatened state is often much less powerful than the United States by whatever crude measure of power

---

[38] Joseph S. Nye, Jr., *Bound to Lead: The Changing Nature of American Power* (New York: Basic Books, 1990), pp. 25–26; Similar definitions come from Max Weber, *Economy and Society: An Outline of Interpretive Sociology* (Berkeley and Los Angeles: University of California Press, 1978), p. 53; and Robert Dahl, *Modern Political Analysis*, 4th ed. (Englewood Cliffs, N.J.: Prentice Hall, 1984), p. 23, and Morgenthau, *Politics among Nations*, pp. 26–27.

[39] See, in general, Morgenthau, *Politics among Nations*; David A. Baldwin, *Paradoxes of Power* (New York: Basil Blackwell, 1989); Gilpin, *War and Change*; on military strength in particular, see Paul Kennedy, *The Rise and Fall of the Great Powers: Economic Change and Military Conflict from 1500 to 2000* (New York: Random House, 1987); Colin Gray, *The Geopolitics of Super Power* (Lexington, Ky.: University Press of Kentucky, 1988), pp. 175–92; on economic growth, Richard Rosecrantz, *The Rise of the Trading State* (New York: Basic Books, 1986); on culture, ideology, and institutions, Nye, *Bound to Lead.*

is used. If simple economic might or military power is sufficient to explain success, there is little reason to work for a more nuanced definition of power.

A single definition of power tends to ignore the relational aspect of power, however. One role that threat may play in the success of internationalization relates to the relative power capabilities of the states involved. David Baldwin, for instance, argues that for a statement about power to have any meaning, it must first specify "who is influencing (or has the capacity to influence) whom."[40] Klaus Knorr concludes that "power is achieved only in particular situations with reference to particular actors."[41] We need to examine the array of states that are targets of internationalization, and the relationship between the power of the target states and that of the sending state. Whatever the components of a general definition of power, these factors are held constant on the sending-state side in this study, since the United States is the only internationalizing state considered. What varies, then, is the power of the target states. We would expect that the weaker the target state, the greater the success at internationalization.

The ability of the target state to resist pressure is likely to depend at least in part on economic characteristics of that state. As Margaret Doxey points out, "vulnerability to economic sanctions is a function of dependence on external supplies of goods or capital and on external markets for domestic products."[42] Efforts have been made to measure economic vulnerability in general, including Gary Clyde Hufbauer, Jeffrey Schott and Kimberly Ann Elliott's measure of the "economic health and stability" of the target states in their study. They conclude that this factor is important in predicting the success of economic sanctions, with the weakest countries the most likely to change their behavior in response to economic pressure.[43]

[40] David A. Baldwin, *Economic Statecraft* (Princeton: Princeton University Press, 1985), p. 20.

[41] Klaus Knorr, "International Economic Leverage and Its Uses," in *Economic Issues and National Security,* ed. Klaus Knorr and Frank N. Trager (Lawrence, Kan.: Regents Press of Kansas, for the National Security Education Program, 1977), p. 109.

[42] Margaret P. Doxey, *International Sanctions in Contemporary Perspective* (New York: St. Martin's Press, 1987), pp. 110–11.

[43] Hufbauer, Schott, and Elliott, *Economic Sanctions Reconsidered,* pp. 46, 97–98; Clark A. Murdock examines the factors that lead to economic vulnerability more broadly, in "Economic Factors as Objects of Security: Economics, Security, and Vulnerability," ed. Knorr and Trager, *Economic Issues and National Security,* pp. 81–96.

Another set of theories about threats looks to the way power is actually used. Power is generally exercised through threats, explicit or implicit. Of particular importance to evaluating theories about threats, since the threat and the use of sanctions are central to efforts by the United States to internationalize its domestic environmental policies, is the vast literature on sanctions in general, and on the conditions or resources seen to make sanctions effective. When we consider sanctions, threats are as important as the actual imposition of sanctions. As any student of deterrence knows, the most successful threat is one that never has to be imposed; therefore, what makes a threat successful is also examined. One of the important issues to emerge from this literature is the role of the costs.

In any effort by one state to persuade another to do something, the costs to both the sending state and the target state of the mechanism of persuasion are likely to be important in influencing the success of the effort, for both sending and target states might bear costs if sanctions are imposed, depending on the structure of the sanctions. Costs to target states come from not being allowed to export their products to the sending state or by not having access to goods or aid from the sending state. Costs to sending states can also arise. Sanctions that prevent exports to target states mean that the companies that would otherwise export to the targets are left without markets to which they might want access. In the case of import sanctions, consumers may be left without foreign goods they want to buy, or industries may be left without required inputs. Any disruption in normal trading relationships is likely to contain costs for some actors. Those who study sanctions have paid particular attention to the issue of costs, though there is disagreement about what the effects of a variety of costs are likely to be.

Measuring costs is itself problematic. Often costs of sanctions are measured by aggregating the economic loss to the sending and the target states and comparing the losses.[44] But as Lisa Martin and others point out, measurement of the cost of sanctions to the sending state relative to the cost to the target state does not provide useful information.[45] Why should it matter if the absolute dollar amount lost by the United States in imposing sanctions on

---

[44] Sidney Weintraub, ed., *Economic Coercion and U.S. Foreign Policy: Implications of Case Studies from the Johnson Administration* (Boulder: Westview Press, 1982), p. 11; Henry Bienen and Robert Gilpin, "Economic Sanctions as a Response to Terrorism," *Journal of Strategic Studies* 3 (May 1980): 89–98.

[45] Martin, *Coercive Cooperation*, p. 56; Baldwin, *Economic Statecraft*, p. 121.

Guyana exceeds the cost to Guyana? Martin argues that what we should measure is the relative cost to the target state of bearing the costs of sanctions versus the costs of changing its behavior.[46] Also seen as important, but difficult to measure, would be the cost to the sending state of imposing sanctions versus the cost of some other action taken to convince the target state to change its behavior.[47]

The effects of cost are also disputed. An important disagreement is over the advantages or disadvantages of costs borne by the sending state. Much of the conventional wisdom on the failure of sanctions in general attributes that failure to the costliness of sanctions to the sending state.[48] Nevertheless, some evidence exists that high costs of sanctions to the sending state can contribute to their effectiveness. This advantage of cost is due largely to the evidence it gives of the sending state's commitment to the goal it pursues. Target states, in determining whether to respond to a threat, consider the likelihood of the threat's being acted upon. Martin finds that in situations in which the state initiating sanctions wants the cooperation of others in the sanctioning efforts, higher costs to the leading sender correlate with higher levels of cooperation on sanctions. By bearing high costs, therefore, a sending state shows its commitment to sanctions.[49] In contrast, costs to a sending state may make it less willing to impose threatened sanctions, and the threat of sanctions may therefore be less credible. Costly sanctions, once imposed, may show the ability of the sending state to overcome domestic opposition, but at the stage when those sanctions are threatened, the potential for domestic opposition to derail them may make them a much less persuasive threat.

Another way to look at the effects of the costs of sanctions to the sending state involves ascertaining whether the sending state has the "political will" to implement the threat. For instance, if the sending state threatens to stop export technology to the target state but the technology industries are privately owned, any sanction is likely to hurt them. Target states may

---

[46] Martin, *Coercive Cooperation,* p. 56.

[47] Baldwin, *Economic Statecraft,* p. 121; he argues that we need to consider both the relative costs and the relative effects of different policy tools.

[48] Weintrab, *Economic Coercion,* p. 11; Hufbauer, Schott, and Elliott, *Economic Sanctions Reconsidered,* pp. 75–82.

[49] Martin, *Coercive Cooperation,* p. 56.

consequently question the likelihood that a sending state will impose costs on its own populations and actually follow through with economic threats.[50] Stephen Krasner emphasizes this credibility aspect as well: leaders "must not only look outward toward their international environment, but they must also look inward toward domestic pressure groups."[51]

Empirically almost all the economic threats studied here involve import restrictions. Import restrictions in general are likely to have a lower cost domestically than export restrictions. Part of the question of whether a sending state can marshal the resources it technically possesses to make effective threats has to do with whether it has control over its constituent parts (can it actually prohibit — legally or logistically — the actions it hopes to prevent); part has to do with whether it is willing to bear the costs — political or economic — that would result from marshaling these resources.[52] In these cases, the net cost of imposing threatened import restrictions is likely to be positive to most organized actors within the sending state. Given what we know about how these threats were authorized at Stage I, we should not be astonished that they do not harm the well-organized actors within the sending state. Even if in some cases there may be advantages to the sending state of bearing high costs as a way to demonstrate resolve, that factor is not likely to come into play among the internationalization attempts studied here. In none of them would imposing sanctions result in the United States suffering an appreciably large net economic loss. There are likely to be some widely dispersed costs to consumers if threats are carried out, however. For instance, there may be less competition for tuna sales, so the price of tuna in the United States may be marginally higher. But these costs will accrue in very small increments to a large and unorganized population, and the benefits will be measurably large to a small and well-organized group. Moreover, the dispersed tuna consumers in this example are very unlikely to organize at the stage at which the sanctions are simply threatened and not imposed, which is the stage at which credibility matters most. Within

[50]  Knorr, "International Economic Leverage and Its Uses," p. 111.

[51]  Stephen Krasner, "Domestic Constraints on International Economic Leverage," in Knorr and Trager, *Economic Issues and National Security*, p. 160.

[52]  Hufbauer, Schott, and Elliott point out that when there are costs to the sending state, import restrictions spread the costs more widely and therefore are less likely to attract the ire of a well-organized domestic group, *Economic Sanctions Reconsidered*, p. 66.

the cases examined here there is, however, variation on the extent of benefits to domestic actors resulting from the imposition of sanctions, and on the distribution of these benefits.

If cost to the threatener is the major determinant of whether a threat succeeds, we would expect greater success of internationalization in cases in which the sending state (in this case the United States) benefits most from the imposition of sanctions. It is there that the sending state would be most likely to carry through its threat to impose sanctions; states wanting to avoid sanctions would therefore adopt the regulations they are pressured to adopt. Actual benefit to major domestic actors in the sending state from the imposition of sanctions would make the threat of sanctions credible.

If target states are willing to accede in the above instances and adopt the regulations in question, it is probably in cases in which the costs of bearing the sanctions would be high. Costs to the target state are, not surprisingly, seen to be important to the question of success of sanctions in general. Hufbauer, Schott, and Elliott calculate, though imperfectly,[53] the costs of sanctions to the target states in their study, on the basis of loss due to lack of access to supplies, markets, finance, or aid, and then estimate the multiplier effect on the target state's economy.[54] They conclude that "cases that inflict heavy costs on the target country are generally successful."[55]

Threatening or inflicting high costs can be an effective form of pressure to convince a target state to take action. This aspect of cost is uncontroversial: the greater the threatened costs to the target state, the more likely it is, ceteris paribus, to adopt the policies it is pressured to adopt, and therefore, the more likely internationalization is to be successful. These threat-based explanations emphasize the idea that target states adopt regulations mainly because of the threat from the sending state, which may succeed as a result of the relative power of the states involved or of the costs incurred. Those who rely on these explanations posit a number of different factors —from raw power to the costs to the sending and target state of the imposition of sanctions—that might lead to the effectiveness of the threat.

---

[53] As Martin points out, their scale does not actually reflect intervals, so that the differences between different levels of measurement on their scale do not represent the same level of cost increase. See Martin, *Coercive Cooperation*, pp. 56–57.

[54] Hufbauer, Schott, and Elliott, *Economic Sanctions Reconsidered*, p. 72.

[55] Ibid., p. 101.

Threats, as vehicles for expressing power, are likely to play some sort of role in explaining the success of internationalization. A simple examination of the relative sizes of the states involved in a conflict, no matter what the unit of measure, will show that the most powerful state frequently prevails. This study makes it immediately apparent that success cannot be a characteristic only of the power of the sending state, however, since only one internationalizing state is considered and there is variation in the levels of success. Even across issue areas, if a particular set of resources in the sending state were the determinants of success in internationalization, the United States should achieve the same results in all countries it attempts to persuade to adopt, say, a certain type of fishery policy. A cursory glance at the evidence quickly disproves this hypothesis. There must, therefore, be a relational aspect of power that matters in determining success of internationalization.

Within that relationship it is likely that both the advantages to the sending state and the disadvantages to the target state, should the threat be acted upon, will play a role in successful internationalization. But neither of those versions of the explanation based on threat address what it is that makes a threat costly or beneficial, nor how we could determine a priori whether it will be one or the other.

### Argument: Market Power and Threat Credibility

Students of international relations will not be surprised that power is expected to play a central role in determining which regulations are successfully internationalized and which states adopted them. What is it about these cases that gives a determining role to power? What aspects of power are expected to be important and in what ways?

It would be a mistake to assume that power resources are the same across issue areas. Different resources will provide power to do different types of things. Possession of nuclear weapons may be useful in deterring use of nuclear weapons by other states, or even in preventing attempts at territorial expansion. It is unlikely to be a useful form of power for altering tariff levels or for convincing another state to adopt stricter emissions standards. Likewise, market dominance in tuna is never considered central to U.S. defense strategy and probably will not even contribute to outcomes in the semiconductor trade. Relating the relevant power resources to the goal pursued is likely to be an important aspect of developing a theory about success in internationalization. So although power is likely to be an important part of the explanation for success, it is necessary to determine what types of power will be relevant for internationalizing environmental regulations.

In the first place, the important power resources for these cases are likely to be economic, since it is economic measures that the internationalizing state uses to threaten states it wants to persuade to adopt regulations. But there are a variety of relative economic measures that provide only marginally more subtle predictions than an aggregated power measure does. Relative GDP, for instance, would provide us a blunt measure that would distinguish inadequately among target states, many of which are significantly poorer than the United States. We need to develop a more nuanced definition of the type of relative economic power that is going to be important.

Johan Galtung focuses on vulnerability as something that can give a sending state power over a target state. He examines "concentration" in the target state as the central concept: "The more a country's economy depends on one product, and the more its exports consist of one product, and the more its exports and imports are concentrated on one trade-partner, the more vulnerable is that country."[56] Eileen Crumm argues that the main determinant of the success of externally imposed incentives "lie[s] inside the target state, in the target state's economic capabilities and its political ability to shift costs and benefits."[57] Knorr similarly derives three factors to consider when calculating the amount of leverage one state is likely to be able to have over another due to economic characteristics. First, the sending state must have a high control over the supply of something that the target state values, such as a resource or a market. The degree of control over that resource may reflect the first of Galtung's points — the extent to which the target relies on resources from, or exports to, the sending state. But it implicitly adds an important aspect: whether there are states other than the sending state that could absorb the exports or provide the goods in question

[56] Johan Galtung, "On the Effects of International Economic Sanctions: With Examples from the Case of Rhodesia," *World Politics* 19, no. 3 (1967): 378–416.

[57] Eileen M. Crumm, "The Value of Economic Incentives in International Politics," *Journal of Peace Research* 32, no.3 (1995): 314.

for the target state. Second, in Knorr's formula, the target state must have need of the supply in question. This factor is similar to Galtung's first point—that the economy of the target state must be to a high degree dependent on the export or import in question. Third, the costs to the target state of complying with the demand must be less than the costs of living with the economic harm imposed by the threat.[58]

As do most studies of sanctions,[59] Galtung's focus on vulnerability refers to the economy of the target state as a whole. But his formulation can be appropriated for use in addressing particular sectors of the economy. In the cases examined here it makes sense to look at the vulnerability of the state in terms of the particular economic sector that is threatened. Two factors, following Galtung's logic, are likely to be important. First, how important is that sector to the economy of the state in question? To take the example in the ozone case of U.S. threats to cut off imports of ozone-depleting substances from target states, this first measurement would judge the proportion of the economy of the target state that relies on exporting that substance: How much of the target country's economy depended on ODS exports? Second, how important is the market in the sending state to the target country? What proportion of ODS exports from the target state go to the United States?

Two economic elements are thus likely to give a sending state market power over a target state in the threatened commodity. The first is the extent to which the good in question is an important part of the economy of the target state—what percentage of the target state's exports can be attributed to the good in question? The second is the extent to which the sending state is an important market for (or, in the case of export restrictions, supplier of) the good—what percentage of the target state's exports of the good go to the sending state? What is likely to matter in determining success in internationalization is the extent to which the sending state has market power relative to the target state in the threatened commodity.

This definition at first seems tautological. Sending states might use economic —and specifically market—threats precisely because they have those power

---

[58] Knorr, "International Economic Leverage and Its Uses," p. 103.

[59] See, for example, Hufbauer, Schott, and Elliott, *Economic Sanctions Reconsidered*; Weintraub, *Economic Coercion*; Doxey, *International Sanctions*; Knorr and Trager, *Economic Issues and National Security*.

resources. What good will this sort of definition do us a priori? Presumably, sending states can also be expected to target their economic power at a sector in which the target state is particularly vulnerable. Until we know at what sector an economic threat will be targeted we will not know if the sending state has the relative market power to make such a threat convincing.

Examining the environmental regulations the United States attempts to push internationally, and the methods by which internationalization is pursued at Stage I, demonstrates that internationalization is likely to be attempted only with a particular set of economic resources that can be identified before the attempt is ever made. Behind the predictability of the economic tools used is the coalition of Baptists and bootleggers at Stage I. Presumably, environmentalists would be willing to use a variety of tools to convince target states to adopt environmental regulations. But full-fledged internationalization attempts are most prevalent when industry actors are also involved, and their goals are protectionist rather than environmental.[60] The measures they support are those that will either make the target states bear the same costs as they do for environmental protection or keep out the goods produced by target states that are not subject to those costs. An examination of the domestic environmental regulation therefore indicates what types of measures will be threatened, should an effort be made to internationalize that regulation. At that point, before internationalization is even attempted, we can ascertain whether the relative market power resources exist to predict the success of attempted internationalization. The presence of the Baptists and bootleggers at Stage I determines the way a threat will be made, and thereby influences its chances at success.

Moreover, it is the presence of this very coalition that makes the threat credible. Both sets of actors within the United States gain if the target state adopts the regulation in question, and the industry actors also gain if economic restrictions are imposed on the target state for not adopting the regulation. Target states know that once a threat is made, a segment of the U.S. population will gain if the threat is actually carried out, which would not be the case with the types of sanctions that are more costly for organized domestic actors. Since threats are occasionally authorized in situations in which the original coalition pushing for them has disintegrated, not all of

---

[60] At the very least they are concerned with improving their competitive advantage.

the cases here involve both industry and environmentalists at Stage II, even though most did at Stage I. Legislation passed by one set of Baptists and bootleggers to allow economic threats in one issue area may be used to threaten in another, or the legislation originally passed may put the industry actors out of business so that they no longer gain from imposition of sanctions. At this point, a third factor, the cost to the target state of adopting the regulation, is also likely to play a role, ceteris paribus. A state may be willing to risk a larger loss because of its weak market position if the regulation itself would be costly. Including in an overall discussion of market power the cost to the target state of adopting the regulation will therefore be important.

As an examination of the cases in which the United States attempts to internationalize domestic environmental regulations in the next chapter demonstrates, the greatest success comes from attempts to internationalize domestic environmental agreements for which the sending state has a high degree of relative market power in the threatened resource, with that resource inherently related to the environmental issue over which the regulation is sought.

# 7

## Markets and Power: International Adoption of U.S. Domestic Policies

The United States has attempted to internationalize a variety of domestic regulations relating to endangered species, air pollution, and fishery conservation. Some of these regulations have been adopted internationally, and some have not. Although ascertaining which domestic regulations a state will attempt to convince others to adopt is an important aspect of the process of internationalization, it is only a preliminary step. If we care either about the ability of a particular state to persuade another to regulate, or about the overall condition of the environment and regulation to protect it, it is also important to know to what extent attempts at internationalization are successful. Under what conditions does the United States succeed in convincing other states to adopt environmental standards equivalent to those the United States already has?

We can begin with the set of cases identified in chapters 3 through 5; instances in which the United States formally attempted to internationalize a domestic regulation through unilateral threats and/or multilateral negotiation. There are several general multilateral efforts at internationalization and approximately nine issues[1] in the three issue areas examined in which the United States passed legislation authorizing unilateral actions to internationalize domestic regulations. Additional actions were taken through multilateral negotiation. Within endangered species regulations, the United States worked for internationalization of species regulation in general (with focus on protection of some particularly threatened species), and of regulations specifically for elephants, sea turtles, dolphins, and whales.

---

[1] The actual number depends on whether you define instances by the specific regulation the U.S. attempts to internationalize, the legislation through which it attempts to do so, or the number of countries subject to internationalization attempts.

Within air pollution regulations, the United States worked for internationalization of regulations relating to ozone depletion. Within fisheries regulations, the United States worked for internationalization of regulations relating to several general aspects of fisheries, such as driftnet regulations and regulations within Exclusive Economic Zones, as well as for regulations relating to salmon and to various species of tuna.

## Success of Internationalization

To analyze the conditions under which internationalization attempts are successful it is necessary to know what was actually done with the legislation passed at Stage I to allow the United States to work for internationalization. There are several different ways in which internationalization attempts are implemented, primarily via international negotiation, threats of sanctions, or some combination thereof. Within threats of sanctions there are also two different types of threats, distinguished by whether the decision to threaten or impose sanctions is left to the discretion of some actor (usually the president or someone in the executive branch), or whether the sanctions are threatened or imposed automatically. With the first type of sanction an actor is generally in charge of deciding when to "certify" that a state does not meet a certain environmental standard, and then whether to take action if the state does not adopt the regulation it has been certified as not upholding. The various incarnations of the Pelly Amendment to the Fisherman's Protective Act generally work this way. With the second type of sanction, the one followed by the Marine Mammal Protection Act and also used for sea turtle conservation, states are immediately required to meet the same standards that U.S. actors do. All goods in question must be certified as having been made or harvested in a way that follows the same regulations required of U.S. actors in order to be allowed into the United States. This type of sanction is largely automatic. All goods are kept out unless they are proved to have been produced according to standards required of United States actors. So there is variation in the extent to which legislation allowing the United States to work for internationalization will be automatically used.

In addition, domestic political struggles within the United States about how to implement internationalization attempts often continue even after the internationalizing legislation has been passed. These struggles may impact the extent to which the legislation is actually used to work for internationalization. Even automatic sanctions like those involving protection of dolphins or sea turtles are subject to interpretation about how widely they should be implemented. The interaction between domestic politics and international factors is thus important at this stage.

### Endangered Species—CITES

It is possible to see the negotiation of the Convention on Trade in Endangered Species of Wild Fauna and Flora (CITES) as an instance of successful internationalization. The United States Endangered Species Conservation Act of 1969 directed the government to negotiate a general international agreement to protect endangered species. The negotiations for CITES were convened in Washington, D.C., and the United States played an active role in bringing about the agreement.[2] The United States wrote an initial draft of the convention to be discussed at the United Nations Conference on the Human Environment at Stockholm in 1972. The eventual draft presented for debate was prepared jointly by the United States, the International Union for the Conservation of Nature (IUCN), and Kenya.[3]

The United States has also taken some specific actions to hold CITES signatories, or others that have not signed, to the requirements of the agreement itself. The U.S. Fish and Wildlife Service on 25 September 1986 banned all wildlife imports from Singapore because of its refusal to uphold international protections for endangered species in general. In particular, the U.S. Fish and Wildlife Service was concerned about exports of the endangered pangolin (an anteater-like mammal) from Singapore, when the species was not native to Singapore. Although Singapore was not a member of CITES, the United States faulted it for not providing proper information documenting the origins of the wildlife and the "effects of the export or re-export upon the wild populations of wildlife," as required for members of CITES. U.S. Department of State officials pointed out the implausibility of Singapore, a small, mostly urban, country, exporting a large volume of wildlife, unless it were getting it from somewhere else.[4] U.S. inter-

---

[2] Richard Littell, *Endangered and Other Protected Species: Federal Law and Regulation* (Washington, D.C.: Bureau of National Affairs, 1992), pp. 101–2.

[3] U.S. Congress, Senate, 93rd Congress, Senate, Executive Report 93–14.

[4] Patt Morrison, "U.S. Imposes Wildlife Ban on Singapore," *Los Angeles Times*, 3 October 1986: Part 2, p. 1 (Lexis/Nexis).

nationalization attempts did not require that Singapore join CITES, but that it comply with the major provisions of the treaty nevertheless.

Sanctions affected Singapore's trade in tropical fish, in which it earned twelve million dollars annually with the United States, as well as an additional three to five million dollars income from animal products. Singapore's first secretary for economics indicated the widespread effects of the sanctions: "It is not a big industry but there are many . . . little breeders."[5] In response, Singapore acceded to CITES, effective 1 March 1987. It also undertook what the United States saw as a "good faith effort to meet the requirements," including enacting domestic legislation to ban trade in rhinoceros products. And it agreed to provide documentation for all wildlife products exported or re-exported from Singapore. As a result, the United States removed the prohibitions on importing wildlife from Singapore, effective 1 January 1987.[6] A similar ban was put on the Philippines in 1985 when it refused to document the origin of its animal products. This ban affected the lucrative trade in reptile-hide shoes. After three days of import restrictions, the Philippines agreed to U.S. documentation requirements.[7]

Other states were the targets of U.S. internationalization efforts for failing to uphold CITES regulations pertaining to specific species, trade in which was prohibited for U.S. citizens. The Pelly Amendment was applied to sea turtle conservation, to prevent trade in sea turtles, all species of which are designated as endangered or threatened in the United States. On 20 May 1991 the Interior and Commerce Departments certified Japan for diminishing the effectiveness of international treaties protecting endangered species by trading in endangered Olive Ridley and hawksbill sea turtles. Japan, a member of CITES, had lodged a reservation for sea turtles,[8] arguing that exporting products made with sea turtles was critical to its tortoise shell industry. "Federal officials [had] said they hope[d] to resolve the matter through negotiations" rather than through certification.[9]

Japanese wildlife exports (including fish) to the United States earned nearly $400 million annually at the time of the threatened sanctions. Some in the

[5] Ibid., p.1.

[6] *51 FR 34159; 51 FR 36864; 51 FR 47064.*

[7] Morrison, "U.S. Imposes Wildlife Ban," p. 1.

[8] Doing so meant that it was not bound by that provision of the agreement.

[9] John Lancaster, "Endangered Sea Turtle Seen Jeopardized by Japan," *Washington Post,* 19 January 1991: A3.

tortoise shell industry, which employed more than two thousand workers, called the U.S. action akin to "a second atomic bomb."[10] Japan acknowledged the need to close its tortoise shell industry but had hoped to do so over a period of years. After the certification, while President Bush was deciding whether to impose sanctions, Japan entered into informal negotiations with the United States on the issue. Such negotiations prior to certification had failed to yield results, but certification and the accompanying threat of sanctions created an added impetus. Japan ended its trade in one of the species of concern—Olive Ridley turtles—and announced its commitment to withdraw its reservation for that species. It also announced that it would end trade in hawksbill turtles by the end of the year, and ultimately withdraw its CITES reservation for that species as well. No sanctions were imposed.[11]

The same legislation was used to work for protection of rhinoceroses and tigers under CITES. All species of tigers and five species of rhinoceros are listed as endangered under the U.S. Endangered Species Act and are listed on Appendix I of CITES. U.S. citizens, therefore, may not trade in or kill these species. On 7 September 1993 the secretary of the interior certified both the People's Republic of China and Taiwan under Pelly for diminishing the effectiveness of CITES by trading in tiger and rhino parts. Because of the Appendix I listing, trade in these animals for commercial purposes is prohibited. Taiwan, because it is not a member of the United Nations, is not eligible to join CITES. China is a member but has not effectively controlled trade in these species. U.S. policy aimed to persuade these two countries to implement the CITES trade ban, regardless of membership, and to promote conservation internally as well. The United States discussed publicly the option of certifying Yemen and South Korea, but did not take that action.[12]

[10] Keith Bradsher, "Sea Turtles Put New Friction in U.S.-Japan Trade Quarrels," *New York Times*, 17 May 1991: A1 (Lexis/Nexis).

[11] "Sea Turtle Actitivities in Japan: Message from the President of the United States Transmitting a Report on Certification by the Secretaries of the Interior and Commerce Concerning Activities by Nationals of Japan Engaging in Trade in Sea Turtles That Threatens the Survival of Two Endangered Species and Severely Diminishes the Effectiveness of the Convention on International Trade, pursuant to 22 U.S.C. 1978 (a) (2)," U.S. Congress, House, 1st sess., House Document 102–85, 20 May 1991.

[12] Susan Katz Miller, "Will US Sanctions Save the Rhino?" *New Scientist* 137 (1859), p. 9 (Lexis/Nexis).

China responded immediately upon learning that it was about to be certified by issuing a state decree outlawing the buying, selling, import, or export of tiger bones and rhinoceros horn.[13] Three days before the president was required to decide whether to impose sanctions, China increased its measures against illegal trade in rhinoceros horn and tiger bones.[14] Taiwan set up a task force to shut down illegal trade in these products.[15]

President Clinton reported to Congress on 8 November 1993 that he was not imposing the sanctions permitted under Pelly because China and Taiwan had both made good-faith efforts to stop trade in these species. However, Clinton noted that those efforts had yet to result in a diminished trade, so he formed an interagency task force to provide technical assistance to these states to assist their efforts to prevent illegal wildlife trade, and he authorized the provision of law enforcement assistance to these states. The report to Congress laid out specific actions that could be taken by these states to show progress in terminating illegal wildlife trade, such as consolidating stockpiles of tiger and rhinoceros parts, creating a wildlife conservation unit, and increasing enforcement penalties. He set a deadline of March 1994 for these states to make visible progress on illegal wildlife trade or risk sanctions.[16]

At the same time that the United States certified these two states, the Standing Committee of CITES recommended that action be taken against China and Taiwan for their trade in tiger and rhino products. The organization also demanded that rhino horn stocks in China, Taiwan, Yemen, and South Korea be sealed.[17] At the March 1994 CITES meeting, the CITES parties decided

[13] Tom Kenworthy, "U.S. Pressures China, Taiwan on Animal Trade," *Washington Post,* 10 June 1993: A28 (Lexis/Nexis).

[14] "Clinton Urged to Punish China, Taiwan over Wildlife," *Reuter Library Report,* 4 November 1993 (Lexis/Nexis).

[15] Sue Pleming, "U.N. Body Cites China, Taiwan over Rhino Horn," *Reuters,* 7 September 1993 (Lexis/Nexis).

[16] "Violations Relating to Endangered Species: Message from the President of the United States Transmitting a Report Concerning the People's Republic of China and Taiwan Engaging in Trade of Rhinoceros and Tiger Parts and Products that Diminishes the Effectiveness of the Convention on International Trade in Endangered Species of Wild Fauna and Flora (CITES), pursuant to 22 U.S.C. 1978 (b)," U.S. Congress, House, 103rd Congress, 1st sess., House Document 103-162, 8 November 1993.

[17] Environment: Sanctions against China and Taiwan to Save the Rhino," *Inter Press Service,* 7 September 1993 (Lexis/Nexis).

not to sanction China or Taiwan, but concluded that neither was enforcing CITES regulations.

In April, determining that Taiwan had not made sufficient progress in curbing illegal trade in tiger and rhino parts, President Clinton decided to impose a ban on imports of wildlife from Taiwan. Congress urged Clinton also to sanction China, but he refused, explaining that China had made progress on curbing its illegal international wildlife trade.[18] The major Taiwanese exports affected by the ban were reptile leather shoes and handbags, shell jewelry, frog legs, live aquarium fish, and bird feathers.[19]

Taiwan reacted angrily. Premier Lien Chan said that U.S. threats were "unfair," and argued that Taiwan had expanded its enforcement against illegal wildlife trade. At the same time, the Taiwanese premier directed his concerns to his own country, besieging Taiwanese people not to "sacrifice the country's image for medical cures or gourmet food."[20] Wildlife exports from Taiwan amount to nearly $25 million annually, a fairly small proportion of Taiwan's trade with the U.S.[21] Taiwanese officials indicated that fifty-three companies would be affected by the ban.[22]

After its initial reaction, however, Taiwan redoubled its efforts at stopping the illegal wildlife trade. It agreed to inventory its rhino horns and consolidate its stockpiles, instigate harsher penalties for those convicted of illegal traffic in wildlife products, enforce its wildlife protection laws, and increase its conservation system more generally. The Taiwanese government committed to spending $37 million in its efforts.[23] The United States removed sanctions 29 June 1995, arguing that Taiwan had improved in its enforcement of species protection regulations and engaged in a public education program.[24] In April of 1997 the U.S. secretary of the interior determined

---

[18]  *UPI Wire Report*, 11 April 1994 (Lexis/Nexis).

[19]  *59 FR 22043–45.*

[20]  "PM Lien Terms US Sanctions 'Unfair' and 'Regretful,'" *Central News Agency*, 1 April 1994 (Lexis/Nexis).

[21]  "Panel Wants Trade Sanctions on Taiwan Times," *Reuters*, 7 April 1994 (Lexis/Nexis).

[22]  "Taiwan Regrets US Trade Sanction Decision," *Agence France Presse*, 12 April 1994 (Lexis/Nexis).

[23]  Sophia Wu, "Taiwan to Meet CITES Requirements by September," *Central News Agency*, 21 April 1994 (Lexis/Nexis).

[24]  Simon Beck, "US Recognises Move to End Wildlife Trade," *South China Morning Post*, 13 September 1996: 4 (Lexis/Nexis).

that Taiwan had accomplished sufficient progress in protection of endangered wildlife to warrant removal of the certification.[25] China remains certified. Both Taiwan and China undertook some actions in response to U.S. internationalization attempts, and it is likely that the trade in endangered species diminished at least a bit as a result. China's continued certification and Taiwan's lengthy period under sanctions, however, show their resistance to these regulations.

Overall the fact that most of these species protection measures have been pursued in the context of CITES regulations may help the internationalized regulations to become institutionalized. Trade in endangered species is difficult to monitor, however, and important black markets for these products are likely to continue, especially in Asia.

## Elephants

In addition to the general species regulations discussed above, the U.S. instigated a specific campaign to protect elephants. Under the African Elephant Conservation Act of 1988 the United States banned imports of ivory from the seventy-seven states that were not members of CITES, in January 1989. The other provisions of the act were not put into place, however, because the United States, and then CITES, called for a complete ban on trade in ivory. Once there was a ban on trade in ivory in general, it superseded the initial regulation the United States attempted to internationalize. Refusing to import ivory from any state was about as far as the United States could go, within the same issue area, toward internationalizing the regulation.

In some ways, then, this legislation can be seen as a successful instance of internationalization, since the international organization responsible for regulating international action on endangered species adopted the same ban on ivory that the United States adopted. CITES's adoption of the ban, however, meant that the regulations the United States was attempting to internationalize through this legislation—adherence to the CITES ivory control policies—was replaced by a new international regulation. The United States, though influential, was not solely responsible for the international attempt at worldwide African elephant protection. International protection of elephants has certainly improved, so much so that CITES has now

25  *62 FR 23479-80.*

allowed for a limited resumption of ivory trade from those countries with well-managed elephant populations.

## Sea Turtles

U.S. fishers or hunters are not allowed to harm any species of sea turtle protected under the Endangered Species Act or under CITES. To that end, U.S. shrimp fishers are required to use "turtle excluder devices" (TEDs) on their shrimp nets, to ensure that any sea turtles caught in the nets can escape. Section 609 of Public Law 101-162 attempts to internationalize United States TED regulations. The legislation requires that the United States exclude all imports of shrimp unless certain criteria are met by the exporting states.[26] Under these regulations, the Department of State was to certify each year those states in question that had taken proper steps toward implementing the regulations: in 1991, target states needed simply to provide information about their shrimping activities and plans for sea turtle protection; in 1992, states needed to make a commitment to using TEDs; in 1993, they needed to have TEDs on 30-50 percent of their shrimp trawl nets; and by 1994 they needed to use TEDs on all their shrimp trawl nets.

The Department of State initially determined that "the scope of section 609 is limited to the wider Caribbean/Western Atlantic region," since that area covers the migratory range of the turtles in question. The countries to which the State Department decided to apply the regulations were Mexico, Belize, Guatemala, Honduras, Nicaragua, Costa Rica, Panama, Colombia, Venezuela, Trinidad and Tobago, Guyana, Suriname, French Guiana, and Brazil. The guidelines allowed for the three-year phase-in of turtle conservation programs in these states.

State Department officials initially checked compliance with these guidelines by looking at legislation passed in the countries in question, by inspecting shrimp trawl vessels to check whether TEDs were actually in place, and by examining monitoring and enforcement programs in the states to ensure that there was compliance with local regulations.[27] Certification is done each year at the end of April, and any of the states in question that

---

[26] *56 FR 1051.*

[27] Interview with Bill Gibbons-Fly, State Department, 1 April 1994; *58 FR 9015–17.*

are not certified are subject to an immediate embargo of their shrimp exports beginning on 1 May of that year.[28]

A number of the states in the initial group of fourteen subject to the restrictions met each condition as the regulations were phased in, and were allowed to export shrimp to the United States throughout this period. Belize, Brazil, Colombia, Costa Rica, Guatemala, Guyana, Mexico, Nicaragua, Panama, and Venezuela all acceded to U.S. demands that they require U.S.-level protection of turtles in their shrimp-fishing operations, despite the fact that doing so meant adopting equipment and regulations they had not previously used. Of the original group of fourteen, Mexico, Panama, and Brazil are all major exporters of shrimp to the United States, exporting between $10.2 and $50.6 million in shrimp annually.[29] Mexican shrimp fishers were among the first to adopt the new Turtle Excluder Devices, hoping to avoid the same type of embargo that their tuna-fishing counterparts had suffered for years.[30]

The other four in the initial group resisted internationalization to a greater or lesser extent throughout the duration of U.S. efforts (table 7.1). Suriname was the first to resist; it did not provide information in 1991 on its plan for preventing incidental take of sea turtles, so imports of shrimp from Suriname were embargoed beginning 1 May 1991.[31] Suriname was certified, and the embargo lifted, when it provided the requisite information later in the year.[32] In 1992 it took the required steps, but in 1993 it had not met the requirement to have TEDs on 30 percent of its trawlers and so could not export shrimp to the United States.[33] It continued to hold out on accepting U.S.-style regulations on shrimp trawl nets through 1997, and shrimp exports were thus disallowed during that period. In 1998, for almost the first time since the effort at internationalization began, Suriname

---

[28] Farmed shrimp are exempted from the export restrictions.

[29] Chris Woodyard, "Shrimp Caught in Restrictions, U.S. Squeeze on Imports May Up Prices," *Houston Chronicle*, 2 November 1996:1 (Lexis/Nexis).

[30] Howard LaFranchi, "Shrimp Lovers, Take Note," *Christian Science Monitor*, 29 April 1996: 1 (Lexis/Nexis). These sanctions are discussed in the next section, on dolphin protection.

[31] *Miami Herald*, May 9, 1991: A12 (Lexis/Nexis).

[32] Interview with Bill Gibbons-Fly, State Department, 1 April 1994.

[33] Ibid.; *Miami Herald*, May 6, 1993: A14 (Lexis/Nexis).

met the conditions for certification and was allowed to export shrimp to the United States,[34] but certification was revoked again in 1999.

Honduras initially indicated a willingness to adopt sea turtle protection and was allowed to export shrimp to the United States in 1991 and 1992. But it failed to demonstrate at the beginning of 1993 that it had installed TEDs on 30 percent of its shrimp trawl nets, and its shrimp was therefore embargoed.[35] It provided the requisite evidence once sanctions were already in effect; on 16 July 1993 the Department of State certified Honduras and removed the embargo.[36] Honduras successfully met certification requirements for 1994 and 1995, but in 1996 there was evidence that Honduras was not enforcing its regulations and so it was not initially certified. Once Honduras was able to convince State Department representatives that enforcement was again occurring, the embargo was dropped as of 1 August 1996.[37]

Trinidad and Tobago accepted the initial efforts at U.S.-style regulation in 1991 and 1992, when only information and a general commitment to turtle protection was required. The first year TEDs were required on a proportion of its shrimp trawl nets, 1993, it did not meet the requirement initially, and underwent thirteen days of a shrimp embargo before being certified on 13 May 1993. The following year, when the United States required that TEDs be used on all shrimp trawl nets, Trinidad and Tobago failed to meet this level of regulation, and it was not allowed to export shrimp to the United States. Trinidadian government officials estimated that an indefinite ban could cost the country "millions of dollars," noting that that the country earns $1.2 million annually from shrimp exports to the United States.[38] This prohibition continued until 15 August 1995, when the country provided sufficient information to indicate that it had a sea turtle pro-

---

[34] 63 *FR* 30550–51.

[35] Interview with Bill Gibbons-Fly, State Department, 1 April 1994; *Miami Herald*, May 6, 1993: A14 (Lexis/Nexis).

[36] Public Notice 1838, 58 *FR* 40685. Exports of farm-raised shrimp from Honduras had not been embargoed, but during the embargo they were required to be accompanied by an Exporter's Declaration certifying that they were farm-raised.

[37] Public Notice 2423, 61 *FR* 43395.

[38] "U.S. Shrimp Ban Seen Costly for Trinidadians," *Journal of Commerce*, 1 June 1995: 5A (Lexis/Nexis).

tection program comparable to that in the United States.[39] It was able to demonstrate a level of regulation acceptable to the United States for the following years and was allowed to export its shrimp there until the 1999 certification determination.

French Guiana resisted internationalization efforts the most thoroughly of the initial fourteen target states. In 1991 it provided the required information about its shrimping activities and indicated a willingness to use TEDs and was therefore allowed to export shrimp to the United States. It did not meet any of the subsequent requirements. Shrimp exports from French Guiana have therefore been embargoed since 1992, and it has not adopted U.S.-style turtle protection regulations.

While internationalization efforts directed toward the initial fourteen target states proceeded apace, domestic actors within the United States were working to broaden the internationalization effort. The U.S. environmental organization Earth Island Institute filed suit against the secretaries of state and of commerce in an attempt to extend the turtle protection program to all foreign shrimpers. The U.S. Court of International Commerce ruled in favor of that organization at the end of 1995, and required that as of 1 May 1996 all shrimp-fishing states be certified (under the strictest standard, requiring turtle excluder devices on all shrimp trawl nets immediately and a comparable incidental catch rate for turtles to that of the United States) in order to export shrimp to the United States.[40]

The Department of State certified that "the fishing environment in 23 . . . countries does not pose a threat of the incidental taking of sea turtles," either because they catch shrimp only in cold waters where sea turtles are rare or because they harvest shrimp manually in ways that do not threaten sea turtles.[41] Most of the new countries whose fishing environments did require TEDs to protect sea turtles made the requisite changes in their shrimp-fishing practices either immediately or shortly after their shrimp was embargoed. Ecuador, El Salvador, and Indonesia met the requirements as

---

[39] Public Notice 2240; 60 *FR* 43640–41.

[40] 61 *FR* 17342–44.

[41] Public Notice 2379, 61 *FR* 24998–99. The states in the first category are Argentina, Belgium, Canada, Chile, Denmark, Germany, Iceland, Ireland, the Netherlands, New Zealand, Norway, Russia, Sweden, the United Kingdom, and Uruguay. The states in the second category are the Bahamas, Brunei, the Dominican Republic, Haiti, Jamaica, Oman, Peru, and Sri Lanka.

soon as the United States demanded them. Thailand, China, and Nigeria did not initially meet certification requirements and were therefore not allowed to export shrimp to the United States at the beginning of this period. Thailand, the largest exporter of shrimp to the United States,[42] instituted a regulation for commercial shrimp trawlers to use TEDs six months after the certification deadline.[43] China, the fourth largest exporter of shrimp to the United States,[44] initially did not require TEDs of all its shrimp trawlers, but shortly after U.S. sanctions were imposed it implemented a regulation requiring TEDs in fishing situations with a risk of incidental catch of sea turtles. Nigeria's Ministry of Fisheries passed a regulation requiring all shrimp trawlers operating in Nigeria's waters to use TEDs as of the end of 1996, and the U.S State Department verified that all Nigeria's shrimp trawlers had installed TEDs and the country's officials were enforcing its regulations. Thailand was certified on 8 November, China on 23 December, and Nigeria on the first of January the following year.[45] (Nigeria lost its certification in 1998 when it did not respond to U.S. requests that it allow U.S. inspectors to examine its shrimp-fishing fleet.[46])

Some states that catch shrimp in areas where sea turtles may be have not applied for certification and are therefore not allowed to export shrimp to the United States. States in this category include India, Malaysia, and Bangladesh, but it is difficult to create an exhaustive list since the court ruling expanded the application of the regulations to all shrimp-fishing countries. The State Department has worked with those specific states in efforts to convince them to meet certification requirements, but so far, to no avail.[47]

Meanwhile, some of the states in the initial group of fourteen that had consistently received certification began to run into trouble, as U.S. scrutiny increased. Brazil was not initially certified for the 1996 season because, despite its earlier certification, the United States doubted that it had sufficient turtle-protection measures in place. It adopted requirements that all

---

[42] Woodyard, "Shrimp Caught in Restrictions," p. 1; Most of Thailand's shrimp, however, comes from aquaculture and therefore was not subject to trade restrictions.

[43] Public Notice 2469, 61 *FR* 59482.

[44] Woodyard, "Shrimp Caught in Restrictions," p. 1.

[45] Public Notice 2498, 62 *FR* 4826. Since the certification year runs from May through April, Nigeria's certification is considered part of the same year.

[46] Public Notice 2831, 63 *FR* 30550.

[47] Interview with David Hogan, State Department, 17 May 1999.

**Table 7.1**
Implementation of Shrimp/Sea Turtle Sanctions

| Fishing Year | Allowed to Export Shrimp to the U.S. | Not Allowed to Export Shrimp to the U.S. |
|---|---|---|
| 1991 | Belize<br>Brazil<br>Colombia<br>Costa Rica<br>French Guiana<br>Guatemala<br>Guyana<br>Honduras<br>Mexico<br>Nicaragua<br>Panama<br>Suriname (later)<br>Trinidad and Tobago<br>Venezuela | Suriname (early) |
| 1992 | Belize<br>Brazil<br>Colombia<br>Costa Rica<br>Guatemala<br>Guyana<br>Honduras<br>Mexico<br>Nicaragua<br>Panama<br>Suriname<br>Trinidad and Tobago<br>Venezuela | French Guiana |
| 1993 | Belize<br>Brazil<br>Colombia<br>Costa Rica<br>Guatemala<br>Guyana<br>Honduras (after 7/16)<br>Mexico<br>Nicaragua<br>Panama | French Guiana<br>Honduras[a] (until 7/16)<br>Suriname<br>Trinidad and Tobago<br>(until 5/13) |

**Table 7.1**
(continued)

| Fishing Year | Allowed to Export Shrimp to the U.S. | Not Allowed to Export Shrimp to the U.S. |
|---|---|---|
| | Suriname<br>Trinidad and Tobago<br>(after 5/13)<br>Venezuela | |
| 1994 | Belize<br>Brazil<br>Colombia<br>Costa Rica<br>Guatemala<br>Guyana<br>Honduras<br>Mexico<br>Nicaragua<br>Panama<br>Venezuela | French Guiana<br>Suriname<br>Trinidad and Tobago |
| 1995 | Belize<br>Brazil<br>Colombia<br>Costa Rica<br>Guatemala<br>Guyana<br>Honduras<br>Mexico<br>Nicaragua<br>Panama<br>Trinidad and Tobago<br>(after 8/15)<br>Venezuela | French Guiana<br>Suriname<br>Trinidad and Tobago<br>(until 8/15) |
| 1996 | Belize<br>China (after 12/23)<br>Colombia<br>Costa Rica<br>Ecuador<br>El Salvador<br>Guatemala<br>Guyana<br>Honduras (after 8/1) | Brazil<br>China<br>French Guiana<br>Honduras (until 8/1)<br>Nigeria (until 1/14/97)<br>Suriname<br><br>Thailand (until 11/8) |

**Table 7.1**
(continued)

| Fishing Year | Allowed to Export Shrimp to the U.S. | Not Allowed to Export Shrimp to the U.S. |
|---|---|---|
| | Indonesia<br>Mexico<br>Nicaragua<br>Nigeria (after 1/14/97)<br>Panama<br>Thailand (until 11/8)<br>Trinidad and Tobago<br>Venezuela | |
| 1997[b] | Belize<br>Brazil<br>China<br>Costa Rica<br>Guatemala<br>Guyana<br>Honduras<br>Indonesia<br>Mexico<br>Nicaragua<br>Nigeria<br>Panama<br>Thailand<br>Trinidad and Tobago<br>Venezuela | Colombia<br>Ecuador<br>French Guiana<br>Suriname<br><br>Plus such states as:<br>Australia<br>Bangladesh<br>India<br>Malaysia |
| 1998[c] | Belize<br>China<br>Colombia<br>Costa Rica<br>Ecuador<br>El Salvador<br>Guatemala<br>Guyana<br>Honduras<br>Indonesia<br>Mexico<br>Nicaragua<br>Nigeria (after 7/21)<br>Panama<br>Suriname | Bahamas<br>Brazil[d]<br>Brunei<br>French Guiana<br>Nigeria (until 7/21)<br>Venezuela (until 7/21)<br><br>Plus such states as:<br>Australia<br>Bangladesh<br>India<br>Malaysia |

**Table 7.1**
(continued)

| Fishing Year | Allowed to Export Shrimp to the U.S. | Not Allowed to Export Shrimp to the U.S. |
|---|---|---|
| | Thailand Trinidad and Tobago Venezuela (after 7/21) | |
| 1999[e] | Belize China Colombia Costa Rica (after 5/17) Ecuador El Salvador Guatemala Honduras Indonesia Mexico Nicaragua Panama (after 5/17) Thailand | Bahamas Brazil Brunei Costa Rica (until 5/17) French Guiana Guyana Panama (until 5/17) Suriname Trinidad and Tobago plus such states as: Australia Bangladesh India Malaysia |

*Source: Federal Register* for the years in question; interview with Bill Gibbons-Fly, State Department, 1 April 1994; interview with David Hogan, State Department, 17 May 1999; U.S. Department of State Press Statement "Sea Turtle Conservation and Shrimp Imports," 4 May 1999.

[a] Exports of farm-raised shrimp from Honduras were not embargoed, but they were required to be accompanied by an Exporter's Declaration certifying that they were farm-raised.

[b] After the court case, all states that are not certified can be considered unable to export shrimp to the United States. The ones included on this list are those that were original targets of the legislation or that the State Department has worked particularly to convince to adopt U.S.-style sea-turtle protection. Interview with David Hogan, State Department, 17 May 1999.

[c] A number of states whose fishing environments were determined not to "pose a threat of the incidental taking of sea turtles" were also allowed to export shrimp to the United States. 63 FR 30550-1.

[d] Brazil can export hand-harvested or aquaculture shrimp to the United States, but not shrimp harvested with trawl nets. The same holds for 1999.

[e] These are the initial listings and are likely to change as states take action to gain certification. The increased number of states whose shrimp is denied entry is a likely result of the WTO decision and its aftermath.

of its shrimp trawlers install TEDs, was certified on 2 April 1997, and sanctions were removed.[48] The United States also began to conduct inspections and require evidence that states actually enforced turtle-protection regulations they had on the books. Venezuela, although it had regulations mandating the use of TEDs, could not show that it was enforcing the regulations and was therefore not allowed to export shrimp to the United States for a period in 1998.[49] Brazil ran into the same difficulties. Colombia lost its certification in 1997 because of lack of enforcement of regulations. Ecuador, not in the original group but successfully certified in 1996, lost its certification in 1997 because it could not demonstrate that it was enforcing its TED requirements.[50] As the second largest exporter of shrimp to the United States, Ecuador found sanctions against its ocean-caught shrimp costly, since shrimp is Ecuador's third largest export product.[51]

The Bahamas and Brunei, exempted from the requirements to use TEDs under the initial expansion in 1996, were prohibited from exporting shrimp to the United States in 1998. The U.S. Bureau of Oceans and International Environmental and Scientific Affairs determined that there were not enough shrimp in Bahamian waters for commercial shrimp fishing (and so any exports had to be from somewhere else). It also determined that Brunei did harvest a sufficient amount of shrimp to mandate the use of TEDs and had not provided evidence that it required them.[52] Neither state had exported shrimp to the United States the previous year.

Several other states challenged U.S. efforts at internationalization. Thailand, despite meeting U.S. requirements shortly after being sanctioned in 1996 and thereby able to export shrimp to the United States, joined with Malaysia, India, and Pakistan in challenging the United States in the World Trade Organization (WTO). These states claimed that the widened sanctions were illegal under international trade law. A three-member panel appointed by the WTO issued an interim ruling on 6 April 1998 that the

---

[48] Public Notice 2528, 62 *FR* 19157.

[49] Public Notice 2831, 63 *FR* 30550; Public Notice 2868, 63 *FR* 44499.

[50] Public Notice 2550, 62 *FR* 29759.

[51] "U.S. May Ban Shrimp Imports from Latin American Nations," *Journal of Commerce*, 9 January 1991: 4A (Lexis/Nexis); Woodyard, "Shrimp Caught in Restrictions," p. 1. A large percentage of Ecuador's shrimp is also farmed, but not as high a percentage as that from Thailand.

[52] Public Notice 2831, 63 *FR* 30550.

U.S. regulations were contrary to the prohibition of quantitative restrictions found in Article XI(1) of the General Agreement on Tariffs and Trade (GATT), and were not justified under Article XX exceptions to protect the environment.[53]

The United States appealed the decision, but the WTO appellate body did not accept U.S. policy. The trade body determined that protection of sea turtles was a legitimate goal under the WTO and could in theory be accomplished with trade restrictions, but it also found that the particular U.S. measures were still discriminatory and therefore not allowed.[54] In response the U.S. State Department modified slightly the way it applied the rules. It allowed individual vessels to be certified as "turtle safe" and agreed to accept imports of shrimp caught on those vessels even when their home state was not certified. A group of U.S. environmental organizations— including the Sea Turtle Restoration Project, the Sierra Club, the Humane Society, and the American Society for the Prevention of Cruelty to Animals, brought the issue once more before the U.S. Court of International Trade. In April 1999 the court issued a preliminary finding that the State Department's modification of the policy did not comply with U.S. law.[55] Should the court's final ruling uphold its preliminary assessment, an appeal is likely. If such a modification to the regulation is not permitted, the United States would not be able to comply with the WTO ruling unless Congress passes a change in the original legislation, which has not been seriously considered at this point.

More importantly, the United States led efforts to create an international agreement to protect sea turtles in the shrimp-fishing process. A three-year negotiating process took place, at the behest of the United States and Mexico. The Inter-American Convention for the Protection and Conservation of Sea Turtles was signed on 1 December 1996 by the United

---

[53] World Trade Organization, "United States—Import Prohibition of Certain Shrimp and Shrimp Products," Interim Panel Report, unpublished copy obtained from the Office of the United States Trade Representative, April 1998.

[54] World Trade Organization, "United States—Import Prohibition of Certain Shrimp and Shrimp Products," Report of the Appellate Body, AB-1998-4, 12 October 1998, Frances Williams, "U.S. Appeal on Shrimp Import Ban Rejected," *Financial Times*, 13 October 1998: 10.

[55] Nancy Dunne, "Legal Wrangle Engulfs U.S. Shrimp Dispute," *Financial Times*, 14 April 1999: 5.

States, Brazil, Costa Rica, Nicaragua, Peru, and Venezuela. It requires, among other turtle conservation regulations, that all states party to the agreement install turtle excluder devices on their shrimp-fishing nets and reduce, to the greatest extent possible, incidental sea turtle deaths in the process of shrimp fishing.[56] Although it has not yet entered into force, President Clinton has sent the agreement to the Senate for ratification.

Before this legislation, none of the fourteen states in the Caribbean/Western Atlantic region used TEDs on any of their shrimp nets. Now, most of the states in question have made a commitment to do so, have enacted policies requiring the use of TEDs on their shrimp trawlers, and are working to enforce the restrictions laid out by the United States. Most other shrimp-exporting states outside the region have changed their behavior as well. These states gained nothing from the change except the protection of sea turtles (which is of no economic advantage to them) and the avoidance of sanctions. The states that have resisted internationalization of this regulation most strenuously are French Guiana and Suriname; Suriname was allowed only briefly to export shrimp to the United States, in May 1998, and French Guiana is not currently allowed to export shrimp to the United States. Trinidad and Tobago and Honduras have also resisted the level of regulation attempted at various stages, but they have recently taken action that has allowed them intermittent access to the U.S. shrimp market.

When the certification process was widened, China initially resisted, but it has to some extent given in to U.S.-style regulations. Brazil and Nigeria have shown more resistance, creating regulations but either not enforcing them or not allowing U.S. inspectors access to determine whether the regulations are being implemented. More important is the institutionalized resistance demonstrated by Thailand's appeal to the World Trade Organization. Although Thailand was willing to take the interim measures necessary to continue to be able to export shrimp to the United States, it used the tools of international law in an effort to overturn internationalization efforts. Although Thailand has succeeded rhetorically, U.S. policy has not changed, and Thai shrimp trawlers nevertheless continue to uphold U.S. standards.

---

[56] Inter-American Convention for the Protection and Conservation of Sea Turtles, 1996, Article IV(2)(h) and Annex III.

The timing of the policy action by the target states leaves no doubt of the importance of the sanctions in changing their policies. At each stage several states did not meet the new standard in time and their shrimp was embargoed; many of those states did eventually comply with U.S. demands. The initial resistance serves only to underscore that these states would not have changed their policy without U.S. pressure. The effect of the WTO ruling is unclear. To some extent most of the states that would have been influenced by U.S. demands had already changed their behavior by the time of the case. Whether they will continue to use TEDs on their shrimp trawl nets remains to be seen, but from initial evidence it seems likely.

The sea turtle population as a whole is likely to have benefited from this policy, although other threats, primarily habitat loss, consumption of eggs, and international trade for the purpose of traditional medicine, remain. The director of the National Marine Fisheries Service pointed to the increase in nests of Kemp's Ridley turtles on Mexican beaches from 700 in 1985 to more than 3400 in 1998 as an indication of the success of widespread use of turtle excluder devices.[57] The international agreement gives some hope that turtle protection measures may persist once the United States decreases its vigilance, but most target states clearly would prefer not to use TEDs to protect sea turtles.

### Dolphins

The requirement that U.S. fishers limit incidental mortality of dolphins in tuna nets is internationalized through the Marine Mammal Protection Act (MMPA). This legislation denies entry to the United States to all yellowfin tuna from foreign countries unless it meets certain criteria. States intending to export yellowfin tuna to the United States must provide documentation that the fishing operations of the state are "accomplished in a manner which does not result in an incidental rate [of dolphin mortality] in excess of that which results from [United States] fishing operations under these regulations."[58] The requirements for upholding these regulations have changed over the duration of the internationalization effort, as domestic agencies struggled to define both domestic and foreign obligations. The United

---

[57] Rolland A. Schmitten, director, NMFS/NOAA, U.S. Commerce Department, "Fishing 'Green' at Sea," letter to the editor, *Washington Post*, 30 January 1999: A18.
[58] 39 *FR* 32124.

States certifies annually the states that have provided proper documentation and satisfied the requirements of the U.S. legislation; only these states are allowed to export tuna to the United States.

A number of tuna-fishing states changed their behavior with each new U.S. requirement, and were allowed to export tuna to the United States (table 7.2). Bermuda, Canada, the Cayman Islands, Costa Rica, New Zealand, the Netherlands Antilles, and Peru were all able to meet the initial U.S. dolphin protection requirements, and they were never barred from exporting tuna to the United States during the period in which the United States was pushing its regulations internationally. Several states, including El Salvador, and the USSR, resisted dolphin protection initially but were either unimportant targets or gave in without too much fanfare. Ecuador and Spain resisted initially but accepted U.S.-style regulations eventually, and have for the past four years been the only states allowed to export yellowfin tuna to the United States.

Other states, however, were less willing to accept U.S. dolphin protection regulations. Mexico, Venezuela, Vanuatu, Colombia, and Panama all resisted U.S. internationalization attempts for some time; the latter three continued to do so throughout the duration of the internationalization efforts, although they nevertheless changed their behavior in ways that had the effect of increasing the protection of dolphins. Belize has recently fished for tuna in the region without protecting dolphins.

Mexico, one of the major tuna-fishing states in the Eastern Tropical Pacific, was the first important target of MMPA sanctions, though the imposition of these sanctions mostly involved domestic wrangling within the United States rather than a unified effort to impose them. Mexico's early lack of certification (beginning in 1981) is not surprising, given that its tuna was already embargoed under Magnuson Fishery Conservation and Management Act sanctions for seizing U.S. vessels fishing for tuna within an Exclusive Economic Zone the United States did not recognize. Interestingly, Mexico did take steps to receive MMPA certification in 1986, before the MFCMA sanctions were removed. Mexico anticipated that as soon as sanctions were dropped, it would be able export tuna to the United States, which it did from 1987 through 1989. Domestic environmental organizations forced the renewal of an embargo on Mexican tuna in 1990, though, in a court case based initially on technical issues.[59] The U.S. Court of Appeals for the Ninth Circuit, however, granted the Department of Commerce's request that it not be required to impose the sanctions on Mexico and so they were removed on 14 November.[60] The stay issued by that court was overturned on 22 February 1991, and an embargo was reimposed.

Mexico called the embargo a "ploy to sabotage [its] tuna industry," while the Mexican tuna industry organized workshops on how to catch tuna in a way that did not kill dolphins, in hopes of avoiding the tuna embargo.[61] The United States attempted to negotiate an agreement with Mexico and Venezuela in March 1992, in conjunction with the U.S. International Dolphin Conservation Act of 1992, in an effort to settle the dolphin protection issues with these countries. The accord would have prohibited tuna fishing with purse-seine nets for five years, but Mexico's Foreign Ministry rejected the agreement, saying the IDCA would hurt Mexico's tuna industry.[62]

Mexico, in addition to refusing to implement national measures to protect dolphins equivalent to those of the United States, further resisted U.S. regulatory efforts by bringing the case before the dispute settlement process of the GATT. Mexico was joined in its petition by Venezuela. The GATT panel ruled against the United States, finding that the United States violated the GATT principle of national treatment for like goods by discriminating against tuna imports on the basis of how they were caught, and that the United States was not justified in pushing domestic environmental measures unilaterally and extraterritorially.[63] Mexico never called for the panel decision

---

[59] The main issue was the method of calculating adherence to U.S. requirements; the court ruled that such a calculation could not be done with fewer than twelve months of catch data. To achieve certification, Mexico would have to show that its dolphin mortality rate was similar to that of the United States for an entire year. *Earth Island Institute et al. v Mosbacher*, United States District Court for the Northern District of California, No. C-88-1380-TEH, 746 F. Supp. 964, 28 August 1990; *55 FR* 42236.

[60] *Earth Island Institute et. al. v Mosbacher*, United States Court of Appeals for the Ninth Circuit, No. 90-16581, 929 F. 2d 1449, 11 April 1991; *55 FR* 48666.

[61] Katherine Ellison, "Mexican Fleet, U.S. Groups Entangled in 'Tuna War,' " *Orange County Register*, 7 November 1991: A32 (Lexis/Nexis).

[62] Michael Parrish and Juanita Darlin, "Mexico Backs away from Pact on Tuna," *Los Angeles Times*, 4 November 1992: D2 (Lexis/Nexis).

[63] General Agreement on Tariffs and Trade, "Dispute Settlement Panel Report on United States Restrictions on Imports of Tuna [Submitted to the Parties 16 August 1991]," *International Legal Materials* 30 (1991): 1594-1623.

**Table 7.2**
Implementation of Tuna/Dolphin Sanctions

| Fishing Year | Allowed to Export Tuna to the U.S. | Not Allowed to Export Tuna to the U.S. |
|---|---|---|
| 1978 | Bermuda | |
| | Canada | |
| | Congo | |
| | Costa Rica | |
| | Ecuador | |
| | Mexico | |
| | Netherlands Antilles | |
| | New Zealand | |
| | Nicaragua | |
| | Panama | |
| | Senegal | |
| | Spain | |
| | Venezuela | |
| 1979 | Bermuda | |
| | Canada | |
| | Congo | |
| | Costa Rica | |
| | Ecuador | |
| | Korea, Rep. of (after 10/1) | |
| | Mexico | |
| | Netherlands Antilles | |
| | Nicaragua | |
| | New Zealand | |
| | Panama | |
| | Senegal | |
| | Spain | |
| | Venezuela | |
| 1980 | Bermuda | Senegal (after 2/1) |
| | Canada | Congo (after 2/28) |
| | Cayman Isl. (after 11/12) | |
| | Costa Rica | |
| | Ecuador | |
| | Korea, Rep. of | |
| | Mexico | |
| | Netherlands Antilles | |
| | New Zealand | |
| | Nicaragua | |

**Table 7.2**
(continued)

| Fishing Year | Allowed to Export Tuna to the U.S. | Not Allowed to Export Tuna to the U.S. |
|---|---|---|
| | Panama | |
| | Spain | |
| | Venezuela | |
| 1981 | Bermuda | Mexico[a] |
| | Canada | |
| | Cayman Isl. (after 11/12) | |
| | Costa Rica | |
| | Ecuador | |
| | Korea, Rep. of | |
| | Mexico | |
| | Netherlands Antilles | |
| | New Zealand | |
| | Nicaragua | |
| | Panama | |
| | Spain | |
| | Venezuela | |
| 1982 | Bermuda | Mexico |
| | Canada | Peru |
| | Cayman Islands | |
| | Costa Rica | |
| | Ecuador | |
| | El Salvador | |
| | Korea | |
| | Netherlands Antilles | |
| | New Zealand | |
| | Panama | |
| | Venezuela | |
| 1983[b] | Bermuda | Mexico |
| | Canada | Peru |
| | Cayman Islands | USSR |
| | Costa Rica | |
| | Ecuador | |
| | El Salvador | |
| | Netherlands Antilles | |
| | Panama | |
| | Portugal | |
| | Venezuela | |

**Table 7.2**
(continued)

| Fishing Year | Allowed to Export Tuna to the U.S. | Not Allowed to Export Tuna to the U.S. |
|---|---|---|
| 1984 | Bermuda<br>Canada<br>Cayman Islands<br>Costa Rica<br>Ecuador<br>El Salvador<br>Panama<br>Venezuela | Mexico<br>USSR |
| 1985 | Bermuda<br>Canada<br>Cayman Islands<br>Costa Rica<br>Ecuador<br>El Salvador<br>Panama<br>Peru<br>Venezuela | Mexico |
| 1986[c] | Bermuda<br>Canada<br>Cayman Islands<br>Costa Rica<br>Ecuador<br>El Salvador<br>Mexico (after 5/21)<br>Panama<br>Peru<br>Venezuela | Mexico (until 5/21) |
| 1987 | Bermuda<br>Canada<br>Cayman Islands<br>Costa Rica<br>Ecuador<br>Mexico<br>Panama<br>Peru<br>Venezuela | El Salvador<br>USSR |

**Table 7.2**
(continued)

| Fishing Year | Allowed to Export Tuna to the U.S. | Not Allowed to Export Tuna to the U.S. |
|---|---|---|
| 1988 | Bermuda<br>Canada<br>Cayman Islands<br>Costa Rica<br>Ecuador (except 10/15–11/1)<br>Mexico<br>Panama (except 10/15–11/23)<br>Peru<br>Vanuatu (after 11/14)<br>Venezuela<br>(except 10/15–11/23) | Ecuador (10/15–11/1)<br>El Salvador<br>Panama (10/15–11/23)<br>Spain (after 12/14)<br>USSR<br>Vanuatu (until 11/14)<br>Venezuela<br>(10/15–11/23) |
| 1989 | Ecuador<br>El Salvador (after 9/18)<br>Mexico<br>Panama (except 9/7–11/16)<br>Spain (after 2/21)<br>Vanuatu (after 12/11)<br>Venezuela (after 12/11) | El Salvador (until 9/18)<br>Panama (9/7–11/16)<br>Spain (until 2/21) |
| 1990 | Ecuador<br>El Salvador<br>Mexico (except 10/10-11/14)<br>Panama<br>Vanuatu<br>Venezuela | Mexico (10/10–11/14) |
| 1991 | Ecuador<br>Panama<br>Vanuatu (until 3/26)<br>Venezuela (until 3/26) | Mexico (after 2/22)<br>Vanuatu (after 3/26)<br>Venezuela (after 3/26) |
| 1992 | Panama (until 12/22)<br>Vanuatu (after 1/22) | Colombia (after 4/27)<br>Mexico<br>Panama (after 12/22)<br>Vanuatu (until 1/22)<br>Venezuela |
| 1993 | Ecuador<br>Spain[d]<br>Vanuatu | Colombia<br>Mexico<br>Panama<br>Venezuela |

**Table 7.2**
(continued)

| Fishing Year | Allowed to Export Tuna to the U.S. | Not Allowed to Export Tuna to the U.S. |
|---|---|---|
| 1994 | Colombia (5/6–9/27)<br>Ecuador<br>Spain<br>Vanuatu (4/4–9/28) | Colombia (until 5/6 and after 9/28)<br>Mexico<br>Panama<br>Vanuatu (until 4/4 and after 9/28)<br>Venezuela |
| 1995 | Ecuador<br>Spain | Colombia<br>Mexico<br>Panama<br>Vanuatu<br>Venezuela |
| 1996 | Ecuador<br>Spain | Belize (after 10/24)<br>Colombia<br>Mexico<br>Panama<br>Vanuatu<br>Venezuela |
| 1997 | Ecuador<br>Spain | Belize<br>Colombia<br>Mexico<br>Panama<br>Vanuatu<br>Venezuela |
| 1998 | Ecuador<br>Spain | Belize<br>Colombia<br>Mexico<br>Panama<br>Vanuatu<br>Venezuela |
| 1999 | Ecuador<br>Spain | Belize<br>Colombia<br>Mexico<br>Panama<br>Vanuatu<br>Venezuela |

*Source: Federal Register* for the years in question; interview with Allison Routt, National Marine Fisheries Service/National Oceanic and Atmospheric Administration, 21 October 1998.

ª Mexican tuna at this point was already embargoed under Magnuson Fishery Conservation and Management Act sanctions.

ᵇ An embargo of Spanish tuna from the Eastern Tropical Pacific Ocean was in place from 1975 to 1983 under the Tuna Conventions Act (See Tuna Conservation I) so Spain was not able to export tuna during this period for reasons unrelated to dolphin protection.

ᶜ All findings of conformance from the previous year were continued through the end of 1986, to allow adaptation to changing rules.

ᵈ Spain is allowed to export tuna that it catches, but not to re-export tuna to the United States imported from other states (as an intermediary).

to be adopted formally, however. That restraint was almost certainly due to the ongoing negotiations with the United States over the North American Free Trade Agreement (NAFTA), in which Mexico hoped to curry U.S. favor. Although officials differ in their presentation of what commitments were made, it appears that Mexico agreed to forgo further action within the GATT if the United States would modify the implementation of the MMPA's internationalization efforts with respect to Mexican tuna.[64] Mexico did agree to stronger dolphin protection measures in its tuna-fishing process, but the United States held off on removing the tuna embargo. The embargo is estimated to have cost Mexico between $200 and $350 million and is believed to have been responsible for the elimination of more than six thousand jobs.[65]

Mexico has, however, altered its behavior in light of U.S. internationalization efforts. It now fishes for yellowfin tuna with smaller boats, exempted from U.S. restrictions, and uses baitboating and hook-and-line fishing to catch yellowfin tuna for export to the United States.[66] It also participates in the dolphin-monitoring program under the Inter-American Tropical

---

[64] Alessandro Bonanno and Douglas Constance, *Caught in the Net: The Global Tuna Industry, Environmentalism, and the State* (Lawrence: University Press of Kansas, 1996), pp. 197–98.

[65] Robert Collier, "Mexican Fishermen Relieved by Tuna Deal," *San Francisco Chronicle*, 30 July 1997: A8 (Lexis/Nexis); Diego Cevallos, "Fisheries: Mexico Dissatisfied with End of U.S. Tuna Embargo," *Inter Press Service*, 31 July 1997 (Lexis/Nexis).

[66] Interview with Lt. Allison Routt, NMFS/NOAA, 29 April 1999. Tuna boats under 400 short tons are excluded from requirements under the MMPA.

Tuna Association and thereby has observers on 100 percent of its large tuna vessels so that dolphin conservation is ensured.

Venezuela's tuna was first embargoed in 1988. During the summer of 1988 officials from the Inter-American Tropical Tuna Commission inspected the Venezuelan tuna fleet and indicated what changes would need to be made to protect dolphins. At the end of September Venezuela passed a law requiring that Venezuelan tuna fishers live up to U.S. requirements for protecting dolphins.[67] Venezuela was allowed to resume tuna exports to the United States as of the end of November of that year. The United States again imposed sanctions in 1991 because Venezuela was unable to show that its dolphin mortality from tuna fishing was comparable to that of the United States. Those sanctions are still in effect as of this writing.

Venezuela has the second largest tuna-fishing fleet in Latin America. Before the embargo Venezuela exported 70 percent of its tuna to the United States.[68] Venezuelan officials argued that the United States set its standards so as to keep foreign tuna out of the American market and thus protect the U.S. fleet.[69] During the course of the embargo, Venezuela's tuna-fishing fleet has been reduced from thirty-two to eight boats, and the country has lost at least six thousand jobs.[70]

Panama complied sufficiently with U.S. dolphin protection requirements to be allowed to export tuna to the United States until the international-ized regulations were made stricter. Panama was briefly not allowed to export tuna to the United States in 1988, when it had not given the proper evidence of dolphin protection, but that evidence was provided later in the year and the embargo was removed. In 1992 a Panamanian Presidential Decree declared "setting on dolphins" in the course of tuna fishing to be illegal, but Panama's certification was revoked and sanctions declared when an inter-national observer saw dolphin mortality from setting on dolphins by Panamanian fishing vessels.[71] Panama revised its laws to forbid this prac-

[67] "U.S. Might Lift Ban on Venezuelan Tuna," *Journal of Commerce*, 19 October 1988: 4A (Lexis/Nexis).

[68] "LATAM-Fishing: Region Seeks to Avoid Repeat of Tuna Embargo," *Inter Press Service*, 1 September 1997 (Lexis/Nexis).

[69] James Brooke, "America—Environmental Dictator?" *New York Times*, 3 May 1992: C7 (Lexis/Nexis).

[70] "LATAM-Fishing."

[71] 58 FR 3013.

tice, but it has not submitted data to indicate that its incidental dolphin catch rates are equivalent to those of the United States.[72] It is therefore not currently allowed to export yellowfin tuna to the United States.

Colombia did not attempt to export tuna to the United States during most of the early years of U.S. dolphin protection policies. When it first did in 1992, the United States denied certification because of inadequate observer coverage to support the claim that dolphin mortality was low.[73] In 1994 Colombia submitted evidence that its incidental dolphin mortality rate was sufficiently low to meet U.S. standards, and that all of its fishing vessels had international observers. As a result, the United States removed the tuna embargo.[74] The embargo was reimposed later that year, however, when U.S. policies changed to require evidence that tuna fishers never encircled northeastern offshore dolphins. Colombia did not incorporate this provision into its legislation and lost its certification.[75]

Vanuatu's tuna was first embargoed in 1988 when that country did not submit documentation to indicate that it was adhering to U.S. dolphin protection requirements; after a month Vanuatu submitted the required information and the embargo was removed.[76] It was again sanctioned in 1991 under the court-ordered expansion of the embargo on Mexican tuna, but after it supplied evidence that its rate of incidental dolphin mortality was comparable to that of the United States, the United States removed the embargo. In 1994 Vanuatu had brought its overall incidental dolphin mortality rate below that of the U.S. tuna fleet, but it had a mortality rate for coastal spotted dolphins (which must constitute less than 2 percent of incidental dolphin mortality rates under U.S. law) more than three times higher than permitted according to the data it reported to the United States. Vanuatu, however, participated in the IATTC observer program, and that organization provided data that convinced the U.S. Department of Commerce to certify Vanuatu in mid-1994.[77] Later that year, however, the embargo was reimposed when Vanuatu refused to prohibit setting on offshore spotted dolphins, as required by U.S. legislation.

[72] NOAA, NMFS, "Tuna/Dolphin Embargo Status Update," http://swr.ucsd.edu/psd/embargo2.htm, 1 June 1998; date visited: 13 October 1998.

[73] 57 FR 17858.

[74] 59 FR 35911.

[75] 59 FR 65974.

[76] 53 FR 45953.

[77] 59 FR 15655.

Belize also was forbidden to export tuna to the United States beginning in October 1996, when it failed to submit information indicating a dolphin protection program comparable to the U.S. one and was known to have begun fishing for tuna in the Eastern Tropical Pacific.

The real impact of the internationalization effort, however, can be seen in the diminishing number of states that sought certification each year. The internationalization effort changed the shape of the global tuna industry. Before MMPA regulations were pushed internationally it was U.S. fishers who moved their operations to foreign countries.[78] When the United States pushed to internationalize the regulations, the composition of the global tuna industry changed in response. Some states, such as Bermuda, the Cayman Islands, the Republic of Korea, New Zealand, and the Netherlands Antilles, ceased large-scale commercial fishing for tuna in the Eastern Tropical Pacific altogether and thus did not even continue to apply for certification.

Others scaled down the size of their tuna industry. Peru is now considered to have adopted equivalent regulations, because it stopped fishing with boats large enough to harm marine mammals. Ecuador was granted certification in 1995 when it left large-scale tuna fishing and sold its purse-seine vessel to Panama.[79] Many of the states subject to tuna embargoes decreased their number of vessels significantly during this time. One effect of the internationalization effort, therefore, was to change the shape of the international tuna industry. In doing so it did manage to improve international protection of dolphins.

**Secondary Embargo**    As discussed in chapter, 3 U.S. environmental groups successfully sued the Department of Commerce in 1992 to require the embargo of tuna from "intermediary nations," those that import tuna from states that have not been certified by the United States. The United States had already declared a secondary embargo the previous year on four states (Costa Rica, France, Italy, and Japan) who were known to be exporting tuna from the major target states, but the court ordered that all states that export tuna be held to certification requirements. The expansion of the secondary embargo by court order included an additional twenty states

---

[78]  Bonanno and Constance, *Caught in the Net,* p. 183.

[79]  60 FR 10332.

until they could prove that the tuna they exported was dolphin safe. Most of these were quickly able to provide documentation that they did not sell tuna caught by embargoed states. In the interim, though, the secondary embargo was a tool in the effort at internationalization, since it forced the targets either to cease buying tuna from states that did not accept U.S. regulations, or to separately ensure that such tuna did not make its way to the United States. Several tuna-fishing states that had managed to avoid or shake off primary tuna embargoes were then struck with secondary embargoes (table 7.3).

The European Union called for a second dispute settlement procedure within the GATT to evaluate the embargo on tuna from "intermediary nations." The GATT found against the United States once again,[80] and once again agreement not to bring the panel findings to a full vote was reached outside the institution. The secondary embargo on Costa Rica, Italy, and Japan has continued. Of these, only Italy consistently tries to export tuna to the United States.

**International Agreements**    United States action contributed to international negotiation, both within and outside the Inter-American Tropical Tuna Commission. The IATTC increased its dolphin protection role, creating an observer program through which states that participate allow trained observers on their tuna vessels to ensure that their tuna is caught in a way that protects dolphins. The United States was the driving force behind the 1992 La Jolla Agreement, an agreement covering dolphin protection measures between the IATTC and the nonmember states fishing for tuna in the eastern Pacific. This agreement called for the reduction of dolphin mortality by 80 percent, through setting quotas on dolphin mortality for individual vessels and setting up a system of IATTC observers on boats to monitor quota limits. It was signed by Costa Rica, France, Japan, Mexico, Nicaragua, Panama, Spain, the United States, and Vanuatu—all the major tuna-fishing states in the region.[81] This agreement expires in 1999.

In addition, in 1995 twelve states, including the United States, signed the Declaration of Panama, which aims to reduce dolphin mortality in the Pacific

---

[80] General Agreement on Tariffs and Trade, "United States—Restrictions on Imports of Tuna." Report of the Panel DS29/R, June 1994.

[81] James Brooke, "10 Nations Reach Accord on Saving Dolphins," *New York Times*, 12 May 1992: C4.

**Table 7.3**
Implementation of Secondary Tuna Sanctions

| Fishing Year | Not Allowed to Export Tuna to the U.S. |
|---|---|
| 1991 | Costa Rica |
| | France |
| | Italy |
| | Japan |
| 1992[a] | Canada (until 12/10) |
| | Colombia (until 12/10) |
| | Costa Rica |
| | Ecuador (until 8/5) |
| | France (until 12/10) |
| | Indonesia (until 8/5) |
| | Italy |
| | Japan |
| | Korea (until 8/5) |
| | Malaysia (until 12/10) |
| | Marshall Islands (until 8/5) |
| | Netherlands Antilles (until 12/10) |
| | Panama (until 8/5) |
| | Singapore (until 12/10) |
| | Spain |
| | Taiwan (until 8/5) |
| | Thailand (until 8/5) |
| | Trinidad & Tobago (until 8/5) |
| | UK (until 12/10) |
| | Venezuela (until 8/5) |
| 1993 | Costa Rica |
| | Italy |
| | Japan |
| | Spain[b] |
| 1994 | Costa Rica |
| | Italy |
| | Japan |
| 1995 | Costa Rica |
| | Italy |
| | Japan |
| 1996 | Costa Rica |
| | Italy |
| | Japan |

Table 7.3
(continued)

| Fishing Year | Not Allowed to Export Tuna to the U.S. |
|---|---|
| 1997 | Costa Rica |
| | Italy |
| | Japan |
| 1998 | Costa Rica |
| | Italy |
| | Japan |

*Source: Federal Register* for the years in question; interview with Allison Routt, National Marine Fisheries Service/National Oceanic and Atmospheric Administration, 21 October 1998.

a In 1992 extra intermediary states were added after the Earth Island Institute won its lawsuit forcing the United States to expand the list of states considered intermediary tuna exporters; later the definition of intermediary was changed (with the approval of Earth Island Institute), resulting in the removal of some of these states from the list.

b Spain is allowed to export tuna that it catches but not to re-export tuna to the United States imported from other states (as an intermediary).

Ocean and to seek methods of capturing tuna without killing dolphins.[82] This agreement, brokered by environmental organizations, sets limits on dolphin mortality in return for the lifting of the U.S. embargoes on tuna. In light of both domestic and international efforts to implement U.S. dolphin-protection measures, and in the process of implementing the Declaration of Panama, the U.S. Congress passed legislation in August 1997 to allow for the lifting of the MMPA embargo on tuna for those countries that signed, ratified, and implemented all requirements of an agreement based on the Declaration. Tuna from those states that fish in the Eastern Tropical Pacific and have not implemented those requirements would be excluded.[83]

At the same time the United States was involved in negotiations on an agreement to implement the elements of the Panama Declaration. The Agreement on the International Dolphin Conservation Program, negotiated in 1998, entered into force in 1999. It sets out methods for limiting

[82] "Tuna Fishing: Tuna Accord Approval Not Universal," *Europe Environment*, 14 November 1995 (Lexis/Nexis).

[83] Public Law 104-42, 111 Stat. 1122–39, 15 August 1997; "U.S.: Seven Countries Sign Dolphin Protection Agreement," *AA Newsfeed*, 22 May 1998 (Lexis/Nexis).

incidental dolphin deaths in the purse-seine tuna fishery through establishment of quotas, improvement of gear and fishing techniques, training and certification for captains and crew, and an on-board observer program.[84] This agreement may be the process by which the ongoing tuna embargoes are finally ended. Already research carried out by the United States under the domestic implementing legislation for this agreement has produced a ruling that will relax the requirements for tuna to be considered "dolphin-safe."[85]

Success of internationalization of dolphin protection regulations is mixed and difficult to ascertain, but there are indications that U.S. actions have had important effects. Some states immediately and consistently took action to implement dolphin protection measures equivalent to those in the United States. A number of states undertook to follow the regulations (some after sanctions had been imposed) and documented their efforts even when their tuna was embargoed for other reasons. For example, during the years that Mexican tuna was under a Magnuson Fishery Conservation and Management Act (MFCMA) embargo, that state still followed the MMPA regulations and provided the documentation to prove it, although it still could not export its tuna to the United States because of the other sanctions. It was later, when the MMPA restrictions were stricter and Mexico was no longer under MFCMA sanctions, that Mexico ceased to meet MMPA requirements.

Some states resisted following the U.S. policies and a number of them were subject to embargoes of their tuna for extended periods of time. That the United States had to make the embargo provisions stricter over time and had to include secondary embargoes indicates the sanctioning provisions were not effective enough on their own to impose regulations on states that strongly resisted undertaking them. That the tuna sanctions were brought twice before GATT dispute-settlement panels indicates further resistance to internationalization. In both cases the GATT panel decided that the U.S. sanctions were inconsistent with its GATT obligations.[86]

---

[84] *Agreement on the International Dolphin Conservation Program,* 1998.

[85] Constance L. Hays, "Government Loosens Standards on Tuna Deemed 'Dolphin Safe,'" *New York Times,* 30 April 1999: 18.

[86] "United States: Restrictions on Imports of Tuna" Report of the Panel DS21/R, reprinted in General Agreement on Tariffs and Trade, *Basic Instruments and Selected Documents* Supplement No. 39, Protocols, Decisions, Reports 1991–92

Evidence suggests, however, that even the states that did not meet U.S. standards in the 1990s have changed their fishing practices in ways that protect dolphins. Most participate in the IATTC's international observer program, most have signed the recent international agreements, and many are fishing with smaller boats or methods less likely to endanger dolphins. Their unwillingness to meet the specific U.S. requirements for lifting the embargo may be due in part to recalcitrance, in part to their having found alternate tuna markets or their still being able to export some tuna to the United States, and in part to the domestic legal wrangling within the United States that has made it nearly impossible for the United States to remove the embargoes.

It is worth pointing out that U.S. consumer demand probably played a role in successful internationalization. Demand for tuna caught in ways that did not harm dolphins may have influenced the willingness of foreign fishers to meet U.S. requirements, since they would have a harder time selling their tuna otherwise, even without U.S. sanctions. It is unlikely, however, that internationalization success can be attributed to consumer demand alone. The changes in tuna-fishing practices have gone far beyond those required to satisfy simple "dolphin-safe" labeling requirements, and the timing of target-state action coincides neatly with certification and sanctioning decisions. But it was yet another tool used by domestic environmental organizations in an overall internationalization campaign.

Through this process, however, the protection of dolphins has increased. Dolphin deaths have diminished dramatically as a result of new fishing methods and changes in the international tuna fleet. Tuna fishers caused the death of approximately 600,000 dolphins annually in the 1960s, and more than 100,000 annually in the 1980s; by 1998 only 2,000 dolphins were killed in the process of tuna fishing.[87] The United States has been successful overall at convincing other states to protect dolphins and the dolphins are thriving. The use of the IATTC and the creation of new international agreements increase the likelihood that improvements in dolphin protection will

---

and Forty-eighth Session, Geneva, December 1993; "United States—Restrictions on Imports of Tuna" Report of the Panel DS29/R, June 1994.

[87] Hays, "Government Loosens Standards," p. 18; "The Center for Marine Conservation Commends the Clinton Administration," *U.S. Newswire*, 30 May 1998 (Lexis/Nexis).

last, particularly in conjunction with a fundamental change in the international tuna fleet.

## Whales

The whaling regulations that restrict U.S. whalers are, first, the regulations passed by the International Whaling Commission, and, second, those under the Marine Mammal Protection Act. Under the rules passed by the IWC, commercial whaling was legal until 1986 but subject to restrictions on seasons, catch quotas, and equipment regulations. In addition, before and during the moratorium, whaling has been allowed for scientific research, but it was subject to regulations or recommendations by the IWC as a whole or by its scientific committee. U.S. fishers have not been allowed to kill whales since the implementation of the Marine Mammal Protection Act.

The main legislation the United States has used to internationalize commitments to IWC regulations is the Pelly Amendment to the Fisherman's Protective Act. This legislation was passed, as discussed earlier, to internationalize salmon conservation regulations, but it was written in language broad enough to allow it to be used to internationalize commitments to any "international fishery conservation program." Under this legislation, the United States "certifies" that a state is "diminish[ing] the effectiveness" of the international regulation. Once a state is certified, the President has the option to "prohibit the bringing or importation into the United States of fish products," and must report to Congress within sixty days on any action or lack thereof taken pursuant to certification.[88] In addition, a state's certification under the Pelly Amendment also triggers its certification under the Packwood-Magnuson Amendment, which requires the secretary of state to reduce that state's fishing allocation in U.S. waters by at least 50 percent. If certification continues for a year without the target state's accepting the regulation in question, no fishing rights can be granted to that state within the U.S. EEZ.[89]

No states have actually been sanctioned under the Pelly Amendment for failure to accept whaling regulations. The Pelly Amendment cases thus involve only threats (though the threats have often been sufficient to bring about change in the target state's policy). In addition, because certification is not automatic, the United States can threaten the possibility of certification,

[88] P.L. 92-219 §8.
[89] 16 U.S.C. §182.

a slightly more removed threat that it is more difficult to document completely. The states subject to such threats, however, are often willing to expose them (and they are reported in the media), so it is possible to compile a list of public efforts the United States has undertaken to internationalize IWC regulations that includes certification threats.

The overall success of internationalization of whaling regulations is mixed. In the short term, especially in the early attempts at internationalization, states took steps to join the IWC or to abide by its conservation recommendations largely because of the threat of sanctions (table 7.4). After the United States certified Japan and the USSR in November 1974 for objecting to the 1973 quota on minke whales, they agreed to abide by the quotas. After the 1978 certifications of Chile, Peru, and South Korea for whaling in excess of the IWC quotas, all three took steps to join the IWC. Most importantly, certification or the threat thereof was influential in getting the USSR, Japan, and Norway, the main whaling states at that point, to agree to abide by the moratorium on commercial whaling. The USSR was certified (and sanctioned under Packwood-Magnuson) for two years, after which it announced that it would cease commercial whaling in 1987. Shortly after Norway was certified under Pelly for its objection to the commercial moratorium, it announced that it would suspend commercial whaling after the 1987 whaling season and would reduce its catch in 1987. Japan was not certified for its objection to the moratorium, but a U.S. domestic battle over certification led to pressure on Japan and uncertainty over whether it would be certified and sanctioned. The decision by the secretary of commerce not to certify resulted in a court case that went all the way to the Supreme Court, with the Court ultimately finding that certification was at the discretion of the secretary of commerce.[90] The possibility of certification was nevertheless quite real up to that point, with a lower court ruling that the United States was required to certify Japan.[91] The Japanese delegate to the IWC later described the process by indicating that "Japan had an experience of involuntary withdrawal of objection to the moratorium decision, coerced by a certain nation."[92] Iceland consistently modified its research whaling programs in response to threats by the United States.

---

[90] *Japan Whaling Association v American Cetacean Society*, 478 U.S. 221, 105 S. Ct. 2860 (1986).

[91] "U.S. Ordered to Curb Japan on Whaling," *Los Angeles Times*, 6 August 1985: 2.

[92] International Whaling Commission, 1993 meeting, *Verbatim Report*, p. 198.

**Table 7.4**
Implementation of Whaling Sanctions

| State | Year | Reason | Certified?[a] | Sanctioned? | Result |
|---|---|---|---|---|---|
| Canada | 1996 | Not in IWC; aboriginal whaling against IWC recommendations | Yes (P) | No | No change |
| Chile | 1978 | Not in IWC, whaling in excess of standards | Yes (P) | No | Joined IWC |
| Iceland | 1986 | Export of meat from scientific whaling | No | No | Agreed to export less than 50% of products from scientific whaling |
| | 1987 | Scientific whaling against IWC resolution | No | No | Agreed to follow recommendations of scientific committee |
| | 1988 | Scientific whaling against IWC resolution | No | No | Modified program; reduced number of whales caught |
| Japan | 1974 | Objected to Minke quota, took more than allowed by quota | Yes (P) | No | Agreed to quota |
| | 1984 | Objected to sperm quota and to commercial moratorium | No | No | Withdrew objection; ceased commercial whaling in 1988 |
| | 1988 | Scientific whaling against recommendations | Yes (P) Yes (P-M) | No Yes | No change |

**Table 7.4**
(continued)

| State | Year | Reason | Certified?[a] | Sanctioned? | Result |
|-------|------|--------|-----------|-------------|--------|
| | 1995 | Scientific whaling against recommendations | Yes (P) | No | No change |
| Korea, Rep. of | 1978 | Not in IWC, whaling in excess of IWC quotas | Yes (P) | No | Joined IWC |
| | 1980 | Objected to ban of cold harpoon | No | No | Withdrew objection |
| | 1986 | Scientific whaling | No | No | Suspended scientific whaling until 1987 meeting |
| Norway[b] | 1986 | Objected to moratorium | Yes (P) | No | Agreed to stop commercial whaling after 1987 (didn't remove objection) |
| | 1988 | Scientific whaling, against IWC resolution | No | No | Continued study (US eventually okayed it) |
| | 1990 | Scientific whaling, against IWC resolution | Yes (P) | No | No change |
| | 1992 | Scientific whaling, against IWC resolution | Yes (P) | No | No change |
| | 1993 | Resumption of commercial whaling | Yes (P) | No | No change |
| Peru | 1978 | Not in IWC, whaling in excess of IWC quotas | Yes (P) | No | Joined IWC |
| Spain | 1980 | Objected to fin quota | No | No | Complied with quota |

Table 7.4
(continued)

| State | Year | Reason | Certified?[a] | Sanctioned? | Result |
|---|---|---|---|---|---|
| Taiwan | 1980 | Not in IWC, whaling | No | No | Stopped foreign whaling (August 1981), then banned all whaling |
| USSR | 1974 | Objected to Minke quota, exceeded it | Yes (P) | No | Agreed to quota |
| | 1985 | Objected to Minke quota and to moratorium; took more than quota for 1985 | Yes (P) Yes (P-M) | No Yes | Stopped commercial whaling in 1987 |

*Source: Public Papers of the President* for the years in question; Gene S. Martin Jr. and James W. Brennan, "Enforcing the International Convention for the Regulation of Whaling: The Pelly and Packwood-Magnuson Amendments," *Denver Journal of International Law and Policy* 17, no. 2 (1989): 293–315.

[a] A *P* indicates that a state was certified under the Pelly Amendment. For states certified under Pelly after 1979 Packwood-Magnuson certification, *P-M* is indicated, unless they were exempt from such certification. Public threats that do not result in actual certification are included here as well.

[b] Since Norway does not have a fishing allocation within the Fishery Conservation Zone of the United States, Packwood-Magnuson certification and sanctions did not apply to Norway.

Internationalization was not universally successful, however, as evidenced particularly by Norway's refusal to temper its whaling policies after its initial reluctant agreement to abide by the commercial whaling moratorium in 1987. It later changed its policy and began a commercial whaling program. Canada has not rejoined the IWC, though its policies are unlikely to conflict with those of the organization. Japan's recent refusal to change its scientific whaling plans also points to the inability of the United States to influence whaling practices. In addition, Iceland has withdrawn from the IWC and is therefore no longer subject to its regulations, almost certainly a step backward in terms of internationalization.

Whale populations may have been impacted at least partly by U.S. actions, particularly by early efforts to bring states into the agreement and by later ones to convince states to implement the moratorium. Although most populations of whales are still severely depleted, some stocks, particularly minke whales, have recovered dramatically. U.S. goals have shifted, however. The United States is now one of the major actors blocking a return to commercial whaling, even for those stocks for which it could be done sustainably. If the current U.S. strategy is to prevent any killing of whales, it has not succeeded.

**Ozone Layer Protection**
Aspects of the 1977 Clean Air Act (as well as other separate pieces of legislation) required U.S. industry to phase out use of CFCs in nonessential aerosols. U.S. industries were put on notice that they would not be able to use substances that harmed the ozone layer under existing U.S. legislation.

The United States worked from that time to gain international regulation of ozone-depleting substances. In 1977 the United States hosted the International Conference on the Ozone Layer, which produced the World Plan of Action on the Ozone Layer. The United States also was an important participant in the creation of the 1985 Vienna Convention for the Protection of the Ozone Layer, and the Montreal Protocol that was added to the Vienna Convention in 1987. Throughout the process the United States pushed for deep cuts in the production and use of ozone-depleting substances. It succeeded in helping to create a set of increasingly strong regulations to be undertaken internationally. Internationalization was not simple. Developing states, led by India and China, stayed out of the regulatory process until they were assured in the London Amendments of funding to compensate them for the "incremental costs" of meeting their obligations under any international regulatory system to protect the ozone layer. Nevertheless, a system is now in place in which every state that uses ozone-depleting substances or is likely to has agreed to undertake protection of the ozone layer.[93]

[93] See, generally, Elizabeth R. DeSombre and Joanne Kauffman, "The Montreal Protocol Multilateral Fund: Partial Success Story," in *Institutions for Environmental Aid: Pitfalls and Promise,* ed. Robert O. Keohane and Marc A. Levy (Cambridge, Mass.: MIT Press, 1996).

The measures taken are likely to be long-lasting. The Montreal Protocol has an international organization to oversee its implementation. Moreover, the changes in the way many industries operate make certain ozone-depleting substances no longer useful in those industrial contexts. Once developing countries are required to meet the full obligations of the convention and old technology outlives its usefulness, the demand for ozone-depleting substances will be negligible.

Because ozone-depleting substances can remain in the stratosphere for up to a century, the ozone layer is not yet recovering, and ozone depletion in the late 1990s is more severe than has been previously recorded.[94] The amounts of many of the regulated substances in the stratosphere are beginning to decline, however, leading to the prediction that the ozone layer itself will soon start to improve.[95] This change can be uncontroversially attributed to international action to limit the use of ozone-depleting substances, pushed at least in part by U.S. internationalization efforts.

## Salmon

U.S. fishers under the International Convention for the Northwest Atlantic Fisheries were subject to regulations about the amount of salmon they could catch. The main general legislation used for internationalization of fishery regulations, the Pelly Amendment to the Fisherman's Protective Act, was passed specifically to address the internationalization of high seas salmon-fishing regulations. The mere passage of the Pelly Amendment was sufficient to gain the acquiescence of Denmark, Norway, and West Germany to the salmon quotas they had not previously agreed to uphold. These states upheld subsequent salmon regulations within the agreement, which certainly helped salmon stocks at the time. The Pelly Amendment was not used again to address salmon issues, although protection of salmon is another reason for the prohibition of driftnets, discussed below.

---

[94] Aisling Irwin, "Ozone Layer Vanishing Faster than Predicted," *Daily Telegraph*, 30 April 1999: 11 (Lexis/Nexis).

[95] World Meteorological Organization, "Summary of Scientific Assessment of the Ozone Layer 1998," WMO Press Release, 22 June 1998. http://www.wmo.ch/web/Press/wmo-unep.html, date visited: 2 May 1999.

## Tuna Conservation I—Tuna Conventions Act

U.S. tuna fishers are bound by the decisions of both the Inter-American Tropical Tuna Commission (IATTC), which regulates tuna fishing in the Eastern Tropical Pacific Ocean, and the International Convention for the Conservation of Atlantic Tunas (ICCAT), which regulates tuna fishing in the Atlantic Ocean. The legislation the United States used to work to internationalize these regulations was the Tuna Conventions Act (TCA). The TCA allows for internationalization of tuna-fishing obligations to states that are not members of these organizations and yet fish for tuna in the regulated areas, or that are members and are not following the regulations set out by the commissions. It authorizes the United States to ban exports of tuna taken "by vessels of such countries under circumstances which tend to diminish the effectiveness of the conservation recommendations of the Commission."[96]

Tuna Conventions Act sanctions relating to the IATTC were used against Spain, and Panama was threatened with them implicitly. This use of the TCA sanctions process came in 1975, when U.S. tuna fishers alleged that during the closed season, "vessels of certain IATTC members and non-member countries have engaged in a directed fishery for yellowfin tuna within the [Commission's Yellowfin Regulatory Area (CYRA)] contrary to the recommendations of the IATTC."[97] The director of the National Marine Fisheries Service (NMFS) investigated alleged violations and relevant information about the extent of the cooperation by the states in question with the IATTC. NMFS held a hearing to determine the extent of violations of IATTC recommendations.[98] For only two countries did it find evidence of wrongdoing: Panama and Spain.[99] Panama had, in weeks prior to the official finding, made a commitment through diplomatic channels to "take the necessary steps to ensure effective control over its flag vessels" in enforcing the IATTC's conservation recommendations, so the United States did not impose sanctions. Spain, however, had vessels fishing in the CYRA during the closed season, catching a significant amount of tuna, with no adequate conservation measures in force. So as of 1 November 1975, all yellowfin tuna

[96] 40 *FR* 48159; 16 U.S.C. 955(c).

[97] 40 *FR* 48159.

[98] 40 *FR* 42230.

[99] 40 *FR* 48160.

"whether received directly or indirectly from Spain" was denied entry into the United States.[100] The embargo was ended once the IATTC ceased regulating tuna catches in the Pacific Ocean, a by-product of the expansion of EEZs of member states. On 12 July 1983 the embargo on yellowfin tuna from Spain was removed, because "since January 1, 1980 the IATTC has had no conservation program in effect for member countries in the eastern Pacific Ocean." Spain's activities were no longer "contrary to any effective IATTC recommendations,"[101] even though Spain never accepted the authority of the organization. Spain had, however, stopped purse-seining in the Eastern Tropical Pacific in the intervening years. It re-entered the fishery in the 1990s.

Internationalization was accomplished successfully with respect to Panama, which took action to discipline its domestic fishers after the threat that its actions would be investigated. Spain, not a member of the IATTC, resisted the regulations as long as they were in effect. Although the tuna embargo may have contributed to its departure from the Pacific tuna fishery (which also meant that it no longer violated IATTC regulations), Spain never did agree to the regulations themselves. Pacific tuna by the end of the IATTC regulatory period was seriously depleted. The extent to which bringing Panama under the regulations and forcing Spain out of the fishery influenced the overall health of the tuna stock is unclear, though those actions likely improved the situation somewhat.

Tuna Conventions Act sanctions were also used more recently with respect to regulation under the ICCAT. In 1994, at the behest of the United States and Japan, the ICCAT adopted a Bluefin Tuna Action Plan, designed to identify those states that "diminish the effectiveness" of the agreement by fishing in a regulated area without upholding the regulations. The organization in 1995 identified Belize, Honduras, and Panama as states that were fishing for tuna in the Mediterranean Sea during a closed season. The United States therefore prohibited imports of bluefin tuna from Belize and Honduras as of 20 August 1997, and from Panama as of 1 January 1998.[102] Although it is too soon to know the final results of these sanctions or their impact on tuna stocks, and none of the states has joined the agreement, there are preliminary signs that the internationalization effort may succeed.

[100]  40 *FR* 48160.
[101]  48 *FR* 32832.
[102]  62 *FR* 44422-23.

Panama, which initially denied violating international regulations,[103] indicated that it would take greater steps to prevent ships flying its flag from catching tuna in violation of ICCAT policies.[104] Belize also said it would attempt to curb acts by its fishers that violate ICCAT rules.[105]

In both these cases the U.S. internationalization efforts were undertaken within the context of international agreements. The involvement of ICCAT and other member states in the latter case is likely to increase the effectiveness of those efforts. Fishery conservation is particularly difficult to enforce, however. Although U.S. efforts may improve the situation, it is difficult to guarantee the long-run health of the fishery.

### Tuna Conservation II—International Regulation

United States tuna fishers were subject to regulations from the above-mentioned international commissions, but the management authority of these commissions began to break down once states started declaring two-hundred-mile zones over which they would be individually responsible for fishery regulation. The United States argued that the highly migratory nature of tuna meant that tuna fishing could be adequately regulated only by international agreements. Since U.S. fishers were bound only by tuna regulations passed by international authorities, the United States attempted to internationalize the idea that tuna management could only be carried out internationally. The legislation used to work for this goal was the Magnuson Fishery Conservation and Management Act. This act allows the United States to impose an embargo on importation of fish products from states that do not allow U.S. fishers to fish for tuna when such a denial is "in violation of an applicable fishery agreement . . . or as a consequence of a claim of jurisdiction which is not recognized by the United States."[106]

The United States at various points used MFCMA sanctions against Costa Rica, Canada, Ecuador, Mexico, Papua New Guinea, and the Solomon Islands, all after the states in question seized United States tuna

---

[103] "Panama Denies Violating Tuna Conservation Efforts," *Reuters World Service*, 20 November 1996 (Lexis/Nexis).

[104] Lauren Grant, "Panama to Clamp Down on Tuna Fish Violators," *Reuters Financial Service*, 2 December 1996 (Lexis/Nexis).

[105] Michael Tighe, "Tuna Commission Approves Sanctions against Three Countries," *AP Worldstream*, 29 November 1996 (Lexis/Nexis).

[106] 16 U.S.C. 1825.

vessels within two hundred miles of their coasts (table 7.5). Another state, Micronesia, seized U.S. tuna vessels, but the United States did not impose sanctions because the U.S. vessel owners had fished not only within the two-hundred-mile EEZ but also within the twelve-mile territorial sea. The owners paid a fine for fishing violations and to purchase a license.[107]

Some of these states gave in to U.S. demands for access to tuna fairly quickly. Papua New Guinea seized a U.S. fishing vessel on 28 February 1982, but as soon as it was threatened with U.S. sanctions, it resold the seized vessel to the original owner for 5 percent of its value. U.S. tuna fishers agreed to buy fishing licenses in return for guaranteed access to tuna within that country's waters.[108]

U.S. boats fishing for tuna within the EEZs of the Solomon Islands regularly ignored the prohibition on doing so, since it was unlikely the country would be able to enforce its prohibition. Assisted by a visiting Australian trawler, however, the Solomons managed to confiscate the U.S. trawler *Jeanette Diana*, and attempted to sell it at auction. The United States imposed sanctions on fish from the Solomons and made clear that it would not recognize the legality of a sale of the vessel. Faced with this pressure, all potential bidders avoided buying the ship.[109] The United States removed the sanctions and allowed tuna exports from the Solomon Islands when that country sold the vessel back to its original owner.[110]

---

[107] Earlier penalties against Latin American states for seizing U.S. fishing vessels are not considered explicitly in this set of internationalization efforts because there was not a well-defined regulation the United States was trying to get accepted internationally in those instances. To the extent that they mirror the goals expressed explicitly under the MFCMA sanctions, however, they form a broader part of this internationalization effort.

[108] The sanctions were declared and lifted within several days, before they could even take effect. See William O. McLean and Sompong Sucharitkul, "Fisheries Management and Development in the EEZ: The North, South, and Southwest Pacific Experience," 63 *Notre Dame Law Review* 4 (fall 1998): 528; Jon M. Van Dyke and Carolyn Nicol, "U.S. Tuna Policy: A Reluctant Acceptance of the International Norm," in *Tuna Issues and Perspectives in the Pacific Islands Region,* ed. David J. Doulman (Honolulu, Hawaii: East-West Center, 1987), p. 113; "Court in Papua New Guinea Seizes Fishing Ship from U.S.," *New York Times,* 28 February 1982: 16.

[109] Chris Pritchard, "Tuna Fishing Causes a Row between U.S. and Pacific Islands," *Christian Science Monitor,* 9 October 1984: 16 (Lexis/Nexis).

[110] 49 *FR* 33526, for seizing a U.S. fishing vessel within the Solomon Islands' EEZ; removed, 50 *FR* 15273.

**Table 7.5**
Implementation of Tuna II Sanctions

| State | Date Imposed | Date Removed |
|---|---|---|
| Canada | 8/31/79 | 9/4/80 |
| Costa Rica | 2/16/79 | 8/10/79 |
|  | 2/1/80 | 2/26/82 |
|  | 4/24/86 | 10/10/86 |
| Ecuador | 11/21/80 | 4/19/83 |
| Mexico | 7/14/80 | 8/13/86 |
| Papua New Guinea | 2/28/82 | 3/3/82 |
| Peru | 5/1/79 | 10/17/79 |
|  | 2/22/80 | 4/19/83 |
| Solomon Islands | 8/23/84 | 4/17/85 |

*Source: Federal Register* for the years in question.

Canada seized a total of nineteen U.S. fishing vessels in 1979 within Canada's EEZ. U.S. sanctions on Canadian tuna were imposed, but lifted when Canada and the United States first reached an interim agreement and then signed a Treaty on Pacific Coast Albacore Tuna Vessels and Port Privileges, supporting the United States' requirement for international regulation of tuna fishing.[111] Canada did, however, bring the sanctions up in the GATT dispute settlement process, which considered the case after the sanctions had already been removed.[112] The panel gave a mixed finding, indicating that the sanctions were justified on some grounds, but not on others.[113]

Several other countries held out longer in accepting U.S. access to tuna. For some that meant enduring U.S. sanctions for a number of years.

[111] Imposed 44 *FR* 53118 for seizing nineteen U.S. fishing vessels within Canada's EEZ; removed 45 *FR* 58459; See Van Dyke and Nicol, "U.S. Tuna Policy," p. 111.

[112] General Agreement on Tariffs and Trade, "United States—Prohibition of Tuna and Tuna Products from Canada," *Basic Instruments and Selected Documents* 91 (29th Supplement, 1983), pp. 107–9.

[113] The panel found that the sanctions were not justified under Article XI(2)(c), but that they were not "arbitrary and unjustifiable," at least in part because the United States had taken action against other states. The panel also found that U.S. measures were not a disguised restriction on trade because they were publicly announced. GATT, "United States—Prohibition of Tuna and Tuna Products from Canada," pp. 107–9.

Ecuador seized three fishing vessels in 1980 (and others shortly after the initial seizure), and sanctions imposed that year were not removed until three years later. Ecuador shipped approximately $15 million in tuna annually to the United States at that point and reacted angrily, protesting to the Organization for American States,[114] but did eventually return the fishing vessels and agreed to allow some U.S. access to tuna in its waters.[115]

The embargo on Mexico's tuna began at the end of 1980 when Mexico seized a U.S. vessel fishing for tuna within Mexico's EEZ and then canceled all fishing agreements with the United States.[116] This episode was the longest-running of MFCMA sanctions.[117] After six years of sanctions, Mexico agreed to discuss regional, rather than national, regulation of tuna, which would allow access by U.S. vessels to tuna in Mexican waters. It also reportedly agreed to limit tuna exports to the United States voluntarily, to avoid harming the U.S. tuna industry domestically.[118] It is also likely that the negative effects on the U.S. tuna industry of the embargo itself contributed to its removal. When no longer able to export tuna to the United States (trade previously worth between $16 and $50 million annually and accounting for 3 percent of the U.S. tuna market), Mexico broke into the European market, crowding out U.S. tuna exports to Europe. The U.S. tuna-fishing industry, initially a main supporter of the sanctions, expressed concern about the impacts of the Mexican tuna embargo on the world tuna market. [119]

Two other states held out in the short run by negotiating an end to sanctions, only to have them reimposed when they repeatedly refused to allow U.S. fishing vessels access to tuna in their waters. Peru first impounded a

[114] "Tuna War Escalates between U.S. and Ecuador," *Associated Press* (International News), 17 November 1980 (Lexis/Nexis).

[115] Imposed 45 *FR* 77219; removed 48 *FR* 16798.

[116] 45 *FR* 47562; "Mexico Terminating All Fishing Accords with United States," *New York Times*, 29 December 1980: A1 (Lexis/Nexis).

[117] 51 *FR* 29183.

[118] Arthur Golden, "U.S. Plan to Lift Tuna Ban Told," *San Diego Union Tribune*, 6 September 1985: A1 (Lexis/Nexis); "Mexico Terminating All Fishing Accords," p. A1; Patrick McDonnell, "U.S. Reportedly Will Lift Embargo on Tuna from Mexico," *Los Angeles Times*, 6 August 1986: 3 (Lexis/Nexis).

[119] See Alberto Szekely, "Yellow-Fin Tuna: A Transboundary Resource of the Eastern Pacific." 29 *Natural Resources Journal* 4 (fall 1989): 1061–62; Mark Kurlansky, "U.S., Mexico Talk Fishing Rights as the Tuna Market Ebbs," *San Diego Union-Tribune*, 31 May 1984: C4 (Lexis/Nexis).

U.S. tuna-fishing vessel in February 1979, but after its tuna exports were prohibited, Peru released the vessel to its owner.[120] In November it then seized eight U.S. fishing vessels, which were released after their owners paid "unexpectedly low" fines.[121] Because Peru did not recognize the U.S. principle of international tuna regulation and prohibited U.S. vessels from fishing for tuna within its EEZ, the United States embargoed Peruvian tuna beginning in early 1980. The Peruvian foreign minister registered a protest of the sanctions with the U.S. ambassador to Peru. The United States removed these sanctions in 1983.[122]

Costa Rica seized two U.S. fishing vessels in 1979, endured an embargo on its tuna exports, and returned the boats.[123] It impounded another U.S. vessel in 1980 and was not allowed to export its tuna to the United States until 1982, when it agreed to the ship's return. This broader conflict ended' briefly in 1983 when the United States negotiated the Eastern Pacific Ocean Tuna Fishing Agreement (with Costa Rica, Guatemala, Panama, and Honduras). This interim agreement, also called the San Jose Agreement, would have regulated tuna in the region and allowed the signatories to fish for tuna within one another's EEZs.[124] The agreement never entered into force, however, and in 1986, Costa Rica again seized a U.S. vessel, only to endure restrictions on tuna exports and return the vessel later that year.[125]

In all these cases the sanctions imposed were removed once agreements had been reached to return the seized fishing vessels to their owners, though often that return included an agreement that the owners would buy a license to fish within the two hundred-mile EEZ of the state in question. These boats were often released after the United States agreed to pay a fine, sometimes in the form of a requirement that vessel owners "buy" their boats back for a nominal fee. The United States, however, considered this situation successful internationalization—in the short run—because the target states agreed to allow some access within their EEZs to U.S. tuna vessels.

[120]  44 *FR* 25554; removed 44 *FR* 59985.

[121]  "Tuna Boats Back at Sea," *Washington Post*, 16 November 1979: A16.

[122]  45 *FR*; 48 *FR* 16798.

[123]  44 *FR* 10172; 44 *FR* 47431.

[124]  "Reagan Signs Tuna Law," Washington Dateline *Associated Press*, 4 October 1984 (Lexis/Nexis); "Mexico News Briefs," *U.P.I.*, 4 March 1984 (Lexis/Nexis).

[125]  51 *FR* 15571; 51 *FR* 36504.

Nevertheless, Costa Rica and Peru repeatedly seized U.S. tuna boats; clearly, therefore, the individual acceptance of U.S. demands was not permanent. In the short run, though, most of these states gave in to internationalization efforts by the United States.

In the medium term, some of the states involved attempted to negotiate broader agreements that met the U.S. standards and guaranteed access to U.S. tuna-fishing vessels, under certain conditions. Of these agreements, the only ones that entered into force were the United States/Canada treaty and the Treaty of Fisheries in the Southern Pacific. This latter agreement did serve as a compromise between the U.S. position and that of the Pacific island states: although U.S. vessels were given access to the EEZs of states in the region, the principle of tuna management by the coastal states was accepted, the United States basically bought its access with a grant of aid, and the United States agreed to give up the right to impose sanctions in this region.

More important, though, the United States was fighting a losing battle over the long run with respect to access to tuna. Latin American states began claiming jurisdiction over two-hundred-mile fishery zones beginning in the 1950s, and the 1982 UN Law of the Sea Convention recognized such zones without the distinction that the United States tried to make about "highly migratory species." The United States was the sole state that did not recognize the authority of coastal states to regulate tuna within their EEZs. Finally, in the Fishery Conservation Amendments of 1990, the United States removed the exception for "highly migratory species" from its EEZ declaration,[126] and it thereby ended its effort to use unilateral sanctions to guarantee U.S. fishers access to foreign tuna. Overall, then, this attempt at internationalization can be seen as a failure.

It is hard to determine the impact on tuna of U.S. insistence that it could only be regulated internationally. Few believed the U.S. rhetoric that the policy was for the purpose of tuna conservation, since it was implemented to guarantee U.S. access to tuna. But regulation of tuna within EEZs has fared at least as badly as international regulation did.[127] Neither the U.S. effort nor its failure seems to have had a beneficial effect on conserving tuna.

---

[126] P.L. 101-627, Title I, Section 101(b).

[127] Elizabeth R. DeSombre, "Tuna Fishing and Common Pool Resources," in *Anarchy and the Environment*, ed. J. Samuel Barkin and George Shambaugh (Albany: SUNY Press, 1999), pp. 52, 55, 62–63.

### Antarctic Fishery Resources

The access of U.S. fishers to the krill and finfish fisheries in the southern ocean are limited under the Convention for the Conservation of Antarctic Marine Living Resources. Although the U.S. Antarctic Marine Living Resources Convention Act of 1984 allows the United States to ban the import of Antarctic resources taken in violation of this convention, this legislation has not been used in any attempt at internationalization. The lack of attempted internationalization stems from the collapse of the Soviet Union, which put the most pressure on the resources of that area and concern about whose compliance with the treaty was one of the issues leading to passage of the internationalizing legislation.[128] This issue will therefore not be considered any further under an investigation of success of attempted internationalization.

### Driftnets

U.S. fishers are not allowed to engage in large-scale pelagic driftnetting, due to the MMPA prohibition on incidental harming of marine mammals and to concerns about adverse effects on fisheries from such efficient large-scale fishing. The United States was also behind the 1989 United Nations General Assembly Resolution 44/225 that required an end to large-scale pelagic driftnet fishing by 1992. The main legislation used to internationalize this prohibition is the Pelly Amendment, which was expanded by the Driftnet Impact Monitoring Assessment and Control Act of 1987 to allow a ban on fish imports from states that refuse to ensure "effective enforcement of law, regulations and agreements" about the use of driftnets, including a ban on their use by 1989. Another piece of legislation, the Fishery Conservation Amendments of 1990, allowed Pelly certification for states that "conduct or authorize their nationals to conduct large-scale driftnet fishing beyond the exclusive economic zone of any nation in a manner that diminishes the effectiveness of or is inconsistent with" an international agreement governing large-scale driftnet fishing that the United States upholds.[129]

---

[128] Christopher Joyner, "Toward Management of Common Pool Living Marine Resources: Lessons and Strategies from the Southern Ocean," paper presented at Social Science Research Council and MacArthur Foundation Workshop on Managing Environmental Conflict: A Common Pool Resources Approach, Georgetown University, 4 November 1995.

[129] P.L. 101-627, Title I, section 107; 104 Stat. 4443.

The United States certified Taiwan and South Korea in 1989 for possible sanctions under the Pelly Amendment for failing "to enter into cooperative scientific monitoring and enforcement agreements."[130] During this period, fish imports from Taiwan and South Korea amounted to around $350 million each, annually.[131] After certification Taiwan concluded a driftnet agreement with the United States. President Bush directed the secretaries of state and commerce to continue negotiations with South Korea and deferred applications of sanctions. South Korea did ultimately negotiate an agreement with the United States.

The same two states were certified two years later for violating the terms of the agreements they had negotiated. Taiwan expressed outrage at this repeated certification, calling the refusal of the United States to import fish caught with driftnets "a serious violation of free trade."[132] After certification, however, both Taiwan and South Korea took steps to correct the problems noted by the United States, and sanctions were not imposed. South Korea recalled its driftnet vessels on the high seas to port, and imposed penalties on the captains of the vessels violating the agreement. Taiwan reiterated its policy to end high seas driftnet fishing by 30 June 1992, the date given in the UN resolution.[133]

In the mid-1990s attention turned to Italy, which was still fishing with driftnets in the Mediterranean. The U.S. Departments of Commerce and State refused to certify Italy under the driftnet internationalization legislation, hoping to pursue the matter through diplomatic means. They were forced to certify Italy when a number of U.S. environmental groups, led by the Humane Society and the Sierra Club Legal Defense Fund, won a case on the issue before the U.S. Court of International Trade. At that point, U.S. imports of fishery-related products from Italy totaled approximately $1.2 billion annually. Once certified, Italy said it would "do its best to

[130] "Cooperative Measures on Driftnet Fishing: Communication from the President of the United States," U.S. Congress, House, 101st Congress, 1st sess., House Doc. 101-93, 7 September 1989.

[131] "U.S. Considers Trade Sanctions against Taiwan, South Korea," *Journal of Commerce*, 15 August 1991: 3A.

[132] "Taiwan May Sue U.S.," *Journal of Commerce*, 4 December 1991: 5A.

[133] "Driftnet Fishing Violations of the Republic of Korea and Taiwan: Message from the President of the United States," U.S. Congress, House, 102nd Congress, 1st Session, House Document 102-155, 21 October 1991.

cooperate,"[134] and the secretary of commerce determined as of 21 January 1997 that Italy had ceased driftnet fishing.[135] In 1998, the European Union negotiated an agreement, over the resistance of Italy, that would ban fishing with driftnets altogether.[136]

Internationalization of driftnet regulations was largely successful. Most states had either already banned use of pelagic driftnets by the time U.S. internationalization legislation was enacted or did not resist international regulations banning the use of driftnets on the high seas. It was in the context of a 1987 fishery agreement with Japan that the expansion of a ban on pelagic driftnetting was passed.[137] Although both Taiwan and South Korea were reluctant to negotiate driftnet agreements with the United States and to abide by the UN moratorium on high-seas driftnet fishing, both ultimately negotiated and observed the moratorium after being threatened with sanctions. Italy appears to have reluctantly ended driftnet fishing as well. A related development was the Convention for the Prohibition of Fishing with Long Driftnets in the South Pacific, signed on 24 November 1991. This was a multilateral attempt, not initiated by the United States, to regulate driftnet fishing. Although it is too early to tell, the decrease in high-seas driftnet fishing will likely have a beneficial impact on fishery conservation and marine mammal protection. The continued use of driftnets by many states in their own EEZs, however, tempers the environmental benefits of the internationalization effort.

## Explanations of Success

Across issues and countries there is significant variation in the levels of success the United States has had in internationalizing environmental regulations. Protection of the ozone layer is clearly the most successful of the efforts. Internationalization of measures relating to the protection of sea turtles and dolphins has had a high degree of success generally, with some variation

[134] Thomas W. Lippman, "Italy Faces Cutoff of Exports to U.S.," *Washington Post*, 14 March 1996: A24 (Lexis/Nexis); 61 *FR* 18721–22.

[135] United States Congress, House, 143 *Congressional Record*, 21 January 1997: H242.

[136] "World Briefs: Group Approves Ban on Fishing with Driftnets," *St. Louis Post-Dispatch*, 9 June 1998: A5 (Lexis/Nexis).

[137] P.L. 100-220, Title. I.

in the implementation of the measures across states. Protection of salmon and tuna by means of international agreements improved with internationalization, and a widespread decrease in world use of driftnets on the open ocean has resulted from U.S. efforts. Early efforts to internationalize regulations affecting whaling were successful, but recent efforts have not met the same success the early ones did. Endangered species protection has had mixed, but sometimes positive, results. And efforts to persuade states to regulate tuna in their EEZs only through international agreements had small initial successes but failed completely overall.

The following is a breakdown of incidences of successful internationalization. "States that accept" are those that are the targets of internationalization (and hence did not initially regulate their citizens in the manner in question) that adopt the regulations with little or no resistance. "States that resist" are those that ultimately acquiesce but resist adopting the regulation in a fairly serious way first. This category includes those states that were threatened or sanctioned at least twice in the course of an effort at internationalization, as well as those that actually received sanctions for a period of more than two years. Although this category is subjective, it is important to distinguish between those states that accept internationalization fairly simply and those that resist mightily, even if they all ultimately give in. "States that refuse" are those that do not, for the course of the internationalization attempt, agree to adopt the regulation in question to the satisfaction of the United States. For some cases, like the first tuna-conservation cases and some of the whaling cases, the effort at internationalization ended without the adoption of the regulation by the target state; for other cases, states are included in this category if they continue to resist internationalization.

Examining table 7.6 makes it easier to determine the extent to which any one explanation alone accounts for success of internationalization. In the first place, it becomes clear that, whether or not expectation of success played a role at Stage I, it did not play much of a role at Stage II, for quite a few attempts at internationalization are unsuccessful. The United States cannot simply impose its environmental regulations on others.

So what does explain the conditions under which states will adopt U.S. regulations they had not previously undertaken? The legitimacy of the request, the advantages of the regulations, and the characteristics of the threat made against states that do not accept the regulations have been suggested as factors underlying successful internationalization.

## Legitimacy

Legitimacy appears to be largely unrelated to the level of success the United States has in convincing others to adopt its environmental regulations. If it were important, we would expect target states to adopt environmental regulations when the standards themselves, or the methods of persuasion used, are accepted by a large number of states. We might predict internationalization to be more successful when attempted with issues that have already been regulated by a number of states or that are already the subject of an international agreement. We might expect, for instance, the further internationalization of fishery regulations that are already embodied in international agreements to be more successful than those that are not thus embodied, and regulations governing species that are pronounced to be endangered by the Convention on International Trade in Endangered Species to have a greater chance of successful internationalization than those regulations governing species listed simply as threatened under the U.S. Endangered Species Act. Persuading nonmember states to join the Inter-American Tropical Tuna Commission, the International North Atlantic Fisheries Commission, or to join or agree to the regulations of the International Whaling Commission should be relatively successful; persuading states to protect sea turtles might be more successful than convincing them to protect dolphins; working against the Law of the Sea to persuade states not to regulate tuna nationally within their EEZs would likely be unsuccessful.

Legitimacy of this sort does not appear to play a deciding role in the success of internationalization. Measured by the consensus of the international community on the goals of the agreement, whaling regulations, endangered species protection under CITES, and tuna conservation (through existing conservation agreements) should be the most successful internationally. Although CITES regulations have generally been accepted, U.S. attempts to hold states to decisions of the International Whaling Commission have recently met with limited success. Perhaps the most successful effort at internationalization has been the protection of sea turtles through turtle excluder devices, which was not at the time addressed in any international

**Table 7.6**
Internationalization Success by State and Issue

| Issue | States that Accept | States that Resist | States that Refuse |
|---|---|---|---|
| Endangered Species—CITES | Singapore Philippines | China Taiwan | |
| Elephants | Everyone | | |
| Sea turtles | Belize Brazil China Costa Rica El Salvador Guatemala Guyana Honduras Indonesia Mexico Nicaragua Nigeria Panama Venezuela | Colombia Ecuador Thailand Trinidad and Tobago | French Guiana Suriname |
| Dolphins[a] | Bermuda Canada Cayman Islands Congo Costa Rica Ecuador El Salvador Korea, Rep. of New Zealand Netherlands Antilles Nicaragua Portugal Senegal Spain Venezuela | Mexico Peru USSR | Belize Colombia Panama Vanuatu Venezuela |
| Whales | Chile Iceland[b] Peru Spain Taiwan | Korea, Rep. of USSR | Japan Norway |

**Table 7.6**
(continued)

| Issue | States that Accept | States that Resist | States that Refuse |
|---|---|---|---|
| Ozone layer protection | Everyone, eventually | India China | |
| Salmon conservation | Denmark Norway W. Germany | | |
| Tuna conservation - I | Belize Panama | Honduras | Spain |
| Tuna conservation - II | Canada Ecuador Papua New Guinea Solomon Islands | Costa Rica Mexico Peru | Everyone, eventually |
| Driftnets | | Italy Korea, Rep. of Taiwan | |

[a] Even the states seen as holding out on dolphin protection measures have changed their behavior substantially in ways that protect dolphins.

[b] Iceland is an interesting case. It acquiesced to U.S. requests three times without even being officially threatened, and so appears to be a case of successful internationalization. It withdrew altogether from the IWC in 1993, however, and said that it would consider commerical whaling. The United States never threatened it under the Pelly Amendment for this move, but it could certainly be considered a failure of internationalization of whaling regulations.

agreement, and only led to the negotiation of one after unilateral action had achieved a high degree of success.

In terms of legitimacy of means, we would expect to find those regulations that are pursued through multilateral diplomacy to be more successful than those pursued through unilateral threat. Measured by the extent to which multilateral efforts were used to create international regulation, ozone layer protection and endangered species regulations under CITES should be the ones most successfully internationalized. That explanation holds, to some extent, for ozone layer protection, and to a lesser extent for rhinoceros and tiger conservation under CITES. But neither way of examining legitimacy can explain the differing levels of success among unilaterally pursued goals, nor across target states within an issue area. That we see different reactions from different states within the same issue area indicates that deeper analysis is required.

In addition, it was in the cases of dolphin and sea turtle protection that the U.S. methods of pursuing internationalization were officially found by international bodies to be illegitimate. Although there was some backsliding in the acceptance of U.S. authority to mandate sea turtle protection after the WTO ruling, this policy is still one of the most successfully internationalized of those examined here. A similar point can be made about dolphin protection. International illegitimacy of means has not prevented the effectiveness of those means.

## Self-Interest

There is some evidence that self-interest plays a role in target state willingness to adopt environmental regulations the United States demands, though not a decisive one. A state could adopt a regulation out of self-interest either because of the benefit or because of the relative cost of the regulation. If internationalization succeeds because of the benefits of environmental regulation to the state being persuaded to adopt it, we would expect target states to adopt regulations that address environmental problems they experience. Target states would therefore be more likely to adopt air pollution regulations that address transboundary pollution they experience than to adopt regulations that address transboundary pollution that they do not experience. States would be more likely to adopt regulations to conserve fisheries on which they depend, so fishing states in general might be able to see the long-run advantage of driftnet regulations and salmon conservation. Whale conservation originally was in the interest of whaling states, as long as the idea was to conserve whales for the future of the whaling industry; there would thus be some advantage to whaling states of adopting measures to conserve whale populations.[138] It may be difficult to measure the particular benefit of preserving endangered species, so species protection would have a low predicted level of success, with legislation protecting those species that are the most endangered having the greatest chance of success.

[138] Once the goal changed (as some argue it has) to protection of whales as species with rights of their own, the environmental benefits of whale conservation for whaling states declined sharply. That change is consistent with the decline in effectiveness of U.S. threats on behalf of IWC regulations.

The least useful to the target states of all the regulations the United States attempted to internationalize should be the second set of tuna conservation regulations, forbidding domestic regulation of tuna within EEZs. In general, the amount of environmental advantage to the target state specifically resulting from adopting the regulation in question should determine the willingness of the state to accept the internationalization of the regulation.

This explanation seems counterintuitive—if environmental self-interest accounts for adoption of regulations, the regulations might be adopted without pressure, simply because they are good for the target state. At the logical extension of this argument, target state regulation would not require U.S. action since states would simply adopt regulations that benefit them environmentally. There certainly are instances of regulations the United States has already adopted also being adopted by other states, apparently out of environmental self-interest. In Europe, for example, laws requiring catalytic converters on cars (similar to the same equipment requirements within the United States) are beginning to be implemented, owing to the environmental advantages of such regulations rather than to any pressure from the United States. But states are composite entities. There are domestic tradeoffs, and winners and losers, from any regulation. Even policies that may advantage a state in an aggregate way may not be adopted because of the importance of the sectors of society that are harmed by the regulation. What this study picks up, then, are the contentious internationalization attempts; the ones where the target state had not, for whatever reason, already adopted the regulations. In focusing on contention, then, this study examines active efforts by states to move environmental regulation forward on the international level, rather than simply looking at which environmental regulations states choose to adopt.

Contention in these internationalization attempts does not mean, however, that environmental self-interest is necessarily absent from the decisions of target states about whether to accept the regulations in question. States might have nonenvironmental reasons for not adopting regulations that would nevertheless be environmentally beneficial. The cost or political unpopularity of the regulations, for example, may prevent states from adopting them, even if such restrictions could improve the environment. It makes sense that if a state were pressured to adopt a number of externally suggested environmental regulations, the ones it would most likely adopt would be those that provide the greatest degree of environmental benefit.

It is difficult to compare environmental benefit across environmental issues and countries. Nevertheless, a discussion of the economic or environmental effects that lack of regulation brings can give some insight into the relative advantages to the states in question of the regulation the United States is attempting to internationalize. In addition, there are often multiple states targeted by internationalization attempts on any particular environmental issue. In those instances we would clearly expect the states that would benefit most from adopting the regulation (or those harmed most by the lack of regulation of the issue) to be the ones toward which internationalization attempts would most likely be successfully directed.

Also important in determining the relative advantage of a regulation is how costly it is to a state to adopt it. We would expect states to be more likely to adopt regulations that have a low cost. General acceptance of an obligation to protect endangered species, as well as to protect those species that are not the basis of organized economic activity, should not be particularly costly to target states. Protection of sea turtles and dolphins is more costly, since fishing for shrimp and tuna must be done in less efficient ways or using equipment that has some (though in the case of turtle excluder devices not incredibly high) cost. The cost of adopting regulations on commons issues (like fishing or whaling) is more difficult to measure, because it provides a more direct benefit (as discussed earlier) than does protection of endangered species. But if we consider that when the United States attempts to internationalize a commons-type regulation, the target states are those that are free-riding on the regulation, then their adopting the regulation can have a relatively high short-run cost. How high the cost of internationalization will be will depend on how central the potentially regulated activity is to the economic life of the target state. It is likely that protection of the ozone layer and the accompanying changes in the process of industrialization it requires would be the most costly of the regulations pushed internationally.[139] Also costly would be the tuna regulations pertaining to EEZs, since target states would lose sole access to tuna stocks off their coasts.

---

[139] As it turned out, the developed countries agreed to pay the incremental costs developing countries would incur in protecting the ozone layer, but that would not have been initially clear when internationalization was first pushed, and nothing was done to address the costs to developed countries.

Driftnet regulations are also relatively costly to adopt, because those states using them would suffer enormous fishing-efficiency losses if the regulations are internationalized. Somewhere in the middle of the range of predicted success of internationalization based on the cost of regulation would be whale, salmon, and tuna conservation regulations. Even in the case of the most pro-whaling states, whaling accounts for a small domestic industry, so the cost to the state as a whole of having to limit, or stop, this activity is not high.

Explanations based on self-interest alone, however, do not account for the level of success in general, nor do they predict those states that would choose to accept internationalized regulations. There is some correlation between benefit and success, since ozone layer protection, likely to be the most environmentally advantageous to the participants, has the highest degree of success, and tuna conservation II (the principle that states cannot regulate tuna conservation within their EEZs but must regulate it internationally), the lowest. But even within those cases the states that held out—China, India, and others on ozone layer protection, for instance—were not likely to be less harmed by the environmental consequences of ozone layer depletion than those that immediately accepted international regulation. That some important developing countries refused to accept regulation of ozone-depleting substances until compensated indicates that the environmental benefits of adopting regulations will not always lead to their adoption.[140] Even within species protection or conservation measures, the degree to which a species is endangered or depleted does not determine the level of success. The level of dolphin protection accepted internationally is not much lower than the level of protection of sea turtles accepted, when the latter are clearly acknowledged as more seriously threatened.

The cost to the target state is also a predictor of the willingness of states to adopt internationalized regulations, but not to any great degree in and of itself. One of the most costly regulations for target states to accept would be internationalization of the second type of tuna conservation measures, since states would no longer have control over the tuna resources within two hundred miles of their coasts (which control had been a driving force in the creation of EEZs in the first place). Also somewhat costly are both the current whaling moratorium, for states that depend on whaling, and the dolphin protection regulations that make tuna fishing less efficient.

[140] See DeSombre and Kauffman, "Montreal Protocol Multilateral Fund."

These three measures have enjoyed varied but incomplete levels of success in internationalization. A much less costly regulatory measure is the required use of turtle excluder devices (each of which cost only several hundred dollars, and have a small effect on the efficiency of shrimp fishing), and that measure has been relatively successfully internationalized. And the most costly regulation, protection of the ozone layer, is actually the most successfully internationalized. For developing countries the cost in forgone development opportunities of undertaking protection of the ozone layer can be enormous, yet a number of developing countries were willing to sign the Montreal Protocol even before compensation was included for "incremental costs" incurred in the agreement. That compensation did bring the recalcitrant states into the Montreal Protocol indicates, however, that the cost of implementing an international regulation is likely to play some role in the willingness of states to adopt it, regardless of the benefits they gain from the regulation itself.

Where this explanation falls short, though, is in its inability to differentiate among target state reactions. It is not likely that the adoption of a given regulation would be considerably more costly for the states that resist adoption of the regulation than for those that accept it. That is particularly true in the case of sea turtle protection and even to some extent in that of dolphin protection. The actions required are identical across states, and the costs unlikely to vary widely, but there are differences across target states in adopting or resisting regulations. In addition, the actual financial cost to a state of agreeing to take measures to protect endangered species (rhinoceroses and tigers in particular) is low, yet states have resisted.

Self-interest, it seems, will not determine a state's willingness to adopt a given regulation at the behest of another. But the existence of even a weak relationship between self-interest and adoption seen here may still indicate that demonstrating the benefits of or removing the costs of regulation will certainly not be detrimental to and may even be helpful in convincing states to adopt regulations.

### Threat

Since many U.S. attempts to convince other states to adopt environmental regulations are carried out through threats, it is not surprising to find that some aspects of threats influence the success of internationalization. They could do so because of the power of the state making the threat, the credibility of the threat, or the cost to the target state if the threat is implemented. If the general power resources of the sending state, however defined, are important in determining success of internationalization, we should not need to know any details about the environmental issue or regulation in question. The only thing that matters should be how powerful the sending state is relative to the target state. Since this study examines the actions of only one sending state, this explanation can be evaluated by examining the relative power of the target states. If the explanation has validity, then the less powerful ones should be more likely than the more powerful ones to adopt the regulations the United States pushes. Ecuador and Guyana should be much more likely to accept internationalization than should Japan or any European states. And states should behave identically across issues. If Spain is a target of U.S. internationalization attempts on both fishery and dolphin protection issues, for example, we would expect it to react the same way regardless of the regulation in question.

Since threats that are most credible are likely to be the most effective, we might expect target states to be most likely to adopt the regulations in question when they believe that the sending state will actually carry through with the action it threatens. Processes in which threats are automatically imposed (to protect dolphins or sea turtles) should be more credible than those in which they are discretionary (in support of tuna, salmon, and whales). The extent to which a sanctions process is discretionary, however, correlates only partly with the success of internationalization and cannot explain the differing reactions of states within a given internationalization attempt.

Threats are also credible in cases when imposing threatened sanctions would actually benefit the influential actors in the sending state. If important sectors of society gain economically from imposing trade restrictions (the threat most often used) against another state, the threat to impose them is convincing. U.S. actors might gain when they make products that compete with those that would be barred from importation if not produced under the same level of regulation as U.S. products. Since the economic interest of the segments of the regulated society within the United States is an important determinant of internationalization strategy at Stage I, it would not be surprising to find that these same actors would want the threatened sanctions to be imposed at Stage II. Threats might also be particularly credible when the sending state would lose greatly if the target state does not adopt a regulation or the sending state would gain greatly if the target state were to adopt it. Under these conditions, the sending state would have an incentive to ensure that the target state adopt the regulation. If advantage to the sending state is the only determinant of success, we should see only variation in success across regulations rather than across states—if what matters is the credibility of threats of the sending state that comes from the advantages of imposing sanctions, then all states threatened on a particular issue should respond the same way.

It might seem that all threats should be made as effective as possible and that there would be no variation within this factor in the cases examined. But, as noted earlier, the form of the threats is determined by the specific coalition that pushes for internationalization. The threat is therefore related to the issue in question. Threats pertaining to regulations on tuna fishing, for instance, are likely to require restrictions on imports of tuna (or possibly of fish in general); rarely are restrictions imposed on unrelated sectors such as technology or grain, despite the leverage that might be gained from broadening the threat. Within the different types of issues examined here, though, there is variation in the ability of the sending state to make a credible threat. The amount by which a U.S. domestic industry gains varies, by issue or by target state. In addition, as seen in Stage I, specific pieces of legislation passed for the purpose of internationalizing one type of regulation may be broad enough to be used for a different type of regulation. The actual threat intended to promote internationalization is sometimes backed by a different set of actors with the ability to threaten more or less credibly.

A list of the regulations that the United States would benefit most from successfully internationalizing (or from imposing sanctions for failure to internationalize) would begin with dolphin protection and tuna conservation policies. The large size and influence of the U.S. tuna industry and the onerous nature of the domestic regulations on tuna fishing mean that this industry would gain much from internationalization. Sea turtle protection and ozone layer protection also involve relatively large and influential industries that would benefit from the imposition of sanctions on the industries with which they compete. Here, regulations for protection of endangered species in which the United States does not have a market should not be likely candidates for successful internationalization. Among these, however, whaling legislation might be more likely to be successfully internationalized for reasons of sending-state gain, since the fishery industry in general in the United States would gain from the imposition of sanctions for whaling, and the threats would therefore be credible.

If the cost to the target state of refusing to adopt a regulation is important in determining whether a state will accept U.S. internationalization, the cost of potential retaliation is what matters. If the repercussions of not adopting the regulations are high enough, target states will instead choose to take on the new regulation, thus making internationalization successful.

The threat explanations, in sum, focus on the different elements of a threat that make it likely to be successful in inducing a change in the behavior of the target of that threat: the power of those making the threat, the credibility of that threat, and the cost to the target state of suffering the consequences of refusing to change behavior. It is important to note that explanations based on what threats are likely to succeed make the most sense when there is an explicit threat formulated and conveyed about the consequences of not internationalizing. Sometimes unilateral threats and multilateral diplomacy are pursued at the same time, so threats can be a factor in these cases. In some efforts at internationalization, however, threats are not explicitly made. That is true, for example, in the case of ozone depletion, in which the U.S. Congress did not actually vote on proposed legislation to restrict entry of goods made with ozone-depleting substances, once it became clear that multilateral efforts to regulate them were moving forward. The predictability of the form of sanctions, however, combined with the transparency of the U.S. political system, suggest that potential target states might anticipate the types of sanctions that might be threatened if multilateral action does not succeed.

It is clear that the simple power explanation does not account for success of internationalization. Not only are some internationalization attempts unsuccessful, which would be unexpected since the United States is stronger than most target states by any measure of power, but target states respond differently on different issues. Colombia is one of the states that has refused to adopt dolphin protection measures, but it quickly gave in and accepted protection of sea turtles. Panama agreed to tuna conservation measures but not to dolphin protection. Costa Rica protects dolphins, but was one of the early opponents of U.S.-style international regulation of tuna fishing. In addition, some major international actors, such as Japan, accept some internationalized regulations (such as sea turtle conservation under CITES), while some minor actors, such as Suriname, French Guiana, and Vanuatu refuse to adopt a variety of regulations. In short, power alone, no matter how one defines it, does not determine success of internationalization.

The advantage to the sending state in imposing sanctions also does not accurately predict success of internationalization, although it contributes some useful information. Despite the advantages to U.S. tuna fishers and shrimpers from the imposition of U.S. sanctions, there are states that refuse to adopt protections of dolphins and sea turtles. Conversely, ozone layer protection is more successfully internationalized than this hypothesis would predict. On the whole, though, the regulations most successfully internationalized are those for which sending states would gain from imposing sanctions for lack of internationalization. But such an explanation cannot account for the variation within reactions by target states. In cases where it is advantageous to the United States to impose sanctions and its threat is therefore quite credible, we see a range of action from the threatened states. This element of threat, then, cannot alone explain success of internationalization.

**Market Power**
Market power as a basis for explaining internationalization comes from aspects of the threat-based explanation: for one state to have market power over another, the good already subject to regulation in the sending state must be an important part of the economy of the target state, and the sending state must be an important market for the good in question. If the good is not an important aspect of the economy of the target state, then the fear of not being allowed to export it to the United States is unlikely to cause a change in behavior. Likewise, if the export in question is an important aspect of the target state's economy but the United States is not one of the main markets for the good, the target state is also unlikely to fear sanctions. Whether there are alternate places to which the target state could export the good in question is also likely to make a difference. Part of market power as well comes from the cost to the target state of adopting the regulation, since there may be a trade-off for the state between bearing the cost of

adopting the regulation and bearing the anticipated cost of sanctions for not adopting the regulation.[141]

The effect of the threatened sanctions on the target states helps to fill in the picture of the effectiveness of threats in achieving international adoption of U.S. regulations. This influence is clearest in the case of sea turtle protection (see Appendix A). There is an almost perfect correlation between the propensity for a state to adopt sea turtle protection regulations and the importance of the United States as a market for its shrimp. The states that accepted sea turtle protection most readily are those that send a large portion of their shrimp exports to the United States and for whom shrimp exports to the United States constitute a relatively high proportion of their total export earnings. In Belize between 4 percent and 11 percent of total export earnings annually come from exporting shrimp to the United States. Guyana and Honduras have similar export profiles. Even Venezuela, which adopted sea turtle protection despite shrimp exports to the United States accounting for only 0.18 percent of its total export earnings, sends 55 percent of its shrimp exports to the United States. Its shrimp industry may be small, but within the industry the United States is an important market. The state that has resisted regulations the most vigorously, French Guiana, exports no shrimp to the United States and therefore is not hurt by the sanctions that have been consistently imposed. Shrimp exports to the United States from Trinidad and Tobago, another resistant state, account for 0.04 percent of its total export earnings, and those from Suriname are less than 0.01 percent. It is no surprise that these states have been able to resist pressure. India, part of the expanded internationalization process, has not even applied for certification to export shrimp to the United States because it has been able to export there without certification. By increasing the percentage of its export to the United States of farm-raised and hand-harvested shrimp, both of which are exempt from the sea turtle protection regulations, India has ensured that U.S. trade restrictions "in no way affect

[141] As noted, this measure does not work as well for addressing multilateral negotiation in which there were no explicit threats as it does for explaining the success of internationalization through threatened or imposed economic sanctions. Since there is often an implicit (or considered) threat on the good relating to the regulation in question though, this aspect can be measured to some extent regardless of whether there is an explicit threat.

the imports from India."[142] Not surprisingly, India has not adopted U.S. sea turtle protection policy.

The case for the effectiveness of threats, evaluated in light of their impact on target states, is slightly weaker when we look at whaling sanctions (see Appendix B). The most resistant states, Norway and Japan, export relatively less fish to the United States than do the other states, although fish is still an important part of their export earnings, and much of that goes to the United States. One aspect that may explain Norway's willingness to flaunt IWC regulations is the second aspect of the sanctions, effective in 1979, that automatically cuts a state's fishing allocation within the U.S. Fisheries Conservation Zone for those certified for whaling. Norway was not affected by these sanctions because it does not have a fishing allocation within U.S. waters. There are still some puzzles here, however, that are not explained simply by cost to target state. Iceland depends strongly on fish exports in general, and fish exports to the United States account for the largest percentage of its export earnings of any of the states considered here. Yet it was willing to risk U.S. sanctions in resisting the whaling moratorium and then in pulling out of the IWC altogether.

The dolphin sanctions also indicate the importance of the U.S. market to a state when it decides whether to accept U.S.-pushed regulations. A number of the states for which the United States is a major tuna market complied relatively quickly with U.S. dolphin-protection demands. Ecuador, which resisted internationalization efforts at points with respect to shrimp, relies heavily on the U.S. market for its tuna exports, and acceded relatively quickly to U.S. demands that it protect dolphins.[143] Other states, like Peru and Colombia, exported negligible amounts of yellowfin tuna to the United States and resisted undertaking U.S.-style dolphin protection regulations.[144] Vanuatu, a more recent target of tuna sanctions, exported no tuna to the United States during the period before it was a target of internationalization.

[142] "U.S. Curbs on Shrimp Import Not to Affect India," *The Hindu*, FT Asia Intelligence Wire, 6 May 1999 (Lexis/Nexis).

[143] Ecuador exports on average between $2 and $6 million worth of yellowfin tuna to the United States annually, a small but significant source of its total export earnings. http://www.nmfs.gov/, date visited: 21 April 1999.

[144] Colombia's exports during the period in question ranged from $0 to $170,000 annually, generally on the low end of those estimates; Peru exports no yellowfin tuna to the United States. http://www.nmfs.gov/, date visited: 21 April 1999.

Even states that chose not to comply with all the U.S. requirements, like Mexico, Panama, and Venezuela, acted only when they had found alternate tuna markets. All three states were able to change their behavior in ways that allowed them to export some of their yellowfin to the United States— that caught by small boats or not by purse-seine—while sending the rest of it elsewhere. Mexico cultivated both a domestic market where none had previously existed and broke into the European market.[145] The market power of the United States was important, and it was only states that either had never depended on the U.S. market or were able to get around its power that were willing to hold out against U.S. internationalization efforts.

Even in the drive to internationalize the resistance to EEZs, the importance of the U.S. market to a given state correlates closely with the degree to which that state was willing to hold out, with Mexico as an important exception.[146] Otherwise the states that resisted U.S. authority to fish for tuna within their EEZs were the ones with the lowest reliance on the U.S. market for fish. Costa Rica and Peru, recalcitrant states, earned considerably less than 1 percent of their total export earnings by exporting fish to the United States. Canada, Ecuador, Papua New Guinea, and the Solomon Islands, those that accepted U.S. authority after the first imposition of sanctions, earned from just over 1 percent (Canada) to nearly 12 percent (the Solomon Islands) of their export earnings from fish exports to the U.S.[147]

Market power should be seen as most effective in the internationalization process, though, in the context of a credible threat. Threat credibility in these cases comes from the alignment of Baptists and bootleggers (environmentalists and industry actors) that exists on the issue, since both gain from the imposition of sanctions. Industry actors may in some cases gain even more from imposing sanctions than they would simply from successful internationalization. If target states adopt regulations that cause them to meet the same environmental standards that industries in the sending state need to meet, competition will at least be fair, when it might not have been before. But if these target states do not adopt the regulations, compe-

[145] Interview with Lt. J. Allison Routt, NMFS/NOAA, 29 April 1999.

[146] Mexico has traditionally relied heavily on the U.S. market for tuna, but was already by this point, as earlier indicated, developing a tuna market in Europe.

[147] These figures are derived from the United Nations *Commodity Trade Statistics* for the years before sanctions were imposed.

tition will be restricted altogether, since most of the threatened sanctions involve refusing to allow goods made/caught with processes not allowed for U.S. actors into the United States. Thus, industry actors in the sending state will often gain from carrying out the threat, and at the very least many of them will not suffer. Environmentalists may not gain from the imposition of sanctions, per se, but they are concerned about convincing the target state to adopt the regulation in question. They are therefore generally pleased with measures designed to put pressure on these states to improve their environmental records. The sea turtle internationalization process was broadened and the policies to protect dolphins were imposed explicitly at the court-supported insistence of environmental organizations, and those organizations played similar roles in making potential whale-conservation sanctions credible. The presence of environmental organizations is important to the idea of threat credibility, because without their concern, as discovered at Stage I, the United States is unlikely to push as hard for its goal.

The extent to which threat processes are automatic, rather than discretionary, influences the credibility of the threat, but the character of the threat process may be attributable to the Baptist/bootlegger coalition at Stage I. Note that the two threat processes that are automatically implemented, those to protect sea turtles and dolphins, are the ones with the most serious involvement of both environmental groups and industry actors. It is no surprise that the threat processes created for internationalization of those two issues are stronger than those in cases like salmon or whaling, where the coalition was weaker or less involved. It is also not surprising that whaling states have been much more willing to risk sanctions that are rarely imposed under the discretionary sanctioning process.

Since the cooperation of the two types of actors was found at Stage I to be the most important factor in leading the United States to attempt to internationalize domestic regulations, it may not be immediately apparent that there would be variation on this factor at Stage II. But since the actual threat of sanctions often comes at a different time (and sometimes on a different issue) than the effort to pass legislation to allow the threat, the coalition involved at this second stage may involve a different set of actors than the one at the first stage, and may be stronger or weaker than the original coalition. For example, a Baptist-and-bootlegger coalition was behind the passage of the Pelly Amendment to address salmon conservation issues. When the legislation was then taken by environmentalists and used to address internationalization of regulations pertaining to whaling or to the general protection of endangered species, there were no longer industry actors (bootleggers) who would benefit from the imposition of the threat. Attempted internationalization of these types of regulations has less threat credibility and a concomitantly lower degree of success.

## Conclusions

This formulation explains the pattern of internationalization success the United States experiences. The cost to the target state of bearing potential sanctions, should it not accept the internationalized regulation, certainly influences the prospects for success of internationalization. But examining costs further shows that the market power exercised by the United States over the target state on the issue in question is the important aspect of target-state cost. States like French Guiana that do not rely on U.S. markets for shrimp are unlikely to be vulnerable to U.S. threats. The credibility of the threat is important as well. Sea turtle conservation and dolphin protection have the clearest Baptist-and-bootlegger coalition and relatively high degrees of success. There is still serious resistance to dolphin protection or to sea turtle regulations from some states, but it comes from those over whom the United States does not have market power. Similarly, although we might expect Iceland, Norway, and Japan to agree to whale conservation on the basis of the market power issue alone, threats of whaling sanctions have a low degree of credibility since there is no U.S. whaling industry to benefit from (or to push for) the imposition of sanctions on these states. In this case the U.S. fishing industry would gain from imposition of these sanctions, since the United States does not allow imports of whale products. But it is often the regulated industry itself that calls for the sanctions, and in this instance there is no U.S. whaling industry. Elements of the fishing industry might gain broadly from the imposition of sanctions in the whaling cases, but they do not have the moral authority to push for it (since they are not harmed by the regulations the United States is trying to internationalize and thus would appear simply opportunistic). They also are not organized to pay attention to activity on this issue, because they are not the original bootleggers that worked for internationalization. And no sanctions have actually been imposed for lack of whale conservation. As states stand up to the threats, and sanctions are not imposed, the threats become even less credible.

The importance of environmental support to the credibility of threats is visible as well. Tuna conservation (in opposition to EEZs) is the least successfully internationalized regulation and yet enjoyed among the highest degree of industry support (and threatened a product over which the United States had market power). Interestingly, though, these are the cases with the least environmental support. In the effort to require international tuna regulations in the face of expanding EEZs, the United States attempted to portray its efforts as responses to environmental necessities. Although environmentalists initially supported the idea of international regulation of fishery resources, once it became clear that this move was largely an attempt to preserve the access of U.S. tuna fishers to the most productive fishing grounds, their support waned. Bootleggers without Baptists, no matter the degree of market power, will likely fail.

It seems clear, then, that some elements of threat play the most important role in predicting success of internationalization. The cost to the target state of adopting the regulation has some role as well, as does the benefit to the sending state of imposing sanctions. The most successful internationalization occurs when the sending state has both market power and the credibility to make target states believe that it will actually use that power. What is required, then, is an issue regarding a resource in which the United States has strong market power over the states it hopes to persuade to adopt environmental protection. Both environmental actors, for moral suasion, and industry actors, for credibility of threats, must be present within the United States at the time that the internationalization attempt is made. When these conditions are met, the United States has a high degree of success at convincing other states to adopt its domestic environmental regulations.

# 8

## Baptists and Bootleggers Revisited: Conclusions and Implications

One way that international standards arise is when one state takes the initiative to regulate domestically to address a problem that has international dimensions. Because it cannot address the problem completely on its own, and because of the domestic implications of solo regulation, that state's domestic population pushes for the regulation to be adopted internationally. The regulations that the United States decides to push forward internationally are those for which there is a coalition of environmentalist and industry actors (Baptists and bootleggers, in an analogy from Prohibition) who benefit from increasing the number of actors bound by the regulation. For the environmentalists, internationalization increases the protection of the resource in question, since more states refrain from depleting or polluting it. For industry actors, internationalization avoids a situation in which they suffer competitive disadvantage relative to their foreign competitors who do not otherwise have to bear a costly environmental regulation. Alone, neither of these sets of actors has the ability to influence the U.S. domestic political process to push for internationalization, which can be costly for the U.S. government to pursue. Although they might work for the same goal for different reasons, these actors working together are more likely to succeed in convincing the United States to attempt internationalization. The first part of the story allows us to determine which types of U.S. regulations are likely to be candidates for attempted internationalization.[1]

[1] The presence of a coalition of environmentalists and industry may be a necessary condition for internationalization of domestic environmental regulations, but it is unlikely to be a sufficient condition. At any given time there is probably some industrial segment that will benefit from internationalization of a regulation, so it would also be useful to determine what it is that would make industry and environmental support sufficient to explain attempts at internationalization.

The nature of this environmentalist-industry coalition determines the form that internationalization attempts take as well. Environmentalists want to use, in addition to international diplomacy, other tools that will make internationalization more successful. Because they are concerned about resource depletion or pollution, threats they support tend to be those that attempt to lessen environmental damage. If the environmental problem is the death of dolphins due to tuna fishing, they support measures that will make it less profitable for states that harm dolphins to fish for tuna. If the issue is inadequate protection for sea turtles, they want to end support for the foreign industries that harm these species.

Industry actors have different concerns that lead to similar prescriptions. Because these actors are concerned about the competitiveness effects of the regulations, they often first try to remove the regulation to which they are subject. When that strategy appears unlikely to succeed and they decide to work for internationalization, they do so with an eye toward inconveniencing their competitors rather than toward creating the most effective environmental protection. They therefore push for threats that remove the competitiveness advantage to foreign states of not meeting U.S. environmental standards. If the concern of U.S. tuna fishers is the cost they bear because of the requirement to protect dolphins, they want to make sure that they do not have to compete with tuna fishers who catch tuna in ways that do not include dolphin-protection costs. If U.S. shrimp fishers have to buy and install turtle excluder devices to protect sea turtles, they do not want their shrimp to compete in a market with shrimp caught by fishers who do not have to use these devices.

In both examples, the zone of agreement between these two different types of actors is clear. The United States works to convince other states to adopt the environmental regulation in question, and does so by threatening to exclude imports of competitive goods created in a way that does not comply with that regulation. In that way both U.S. domestic groups are satisfied. In one possible scenario, the rule is successfully internationalized, the environment is more fully protected, and the competitive disadvantage to U.S. firms is removed. In the other, the extent to which target states harm the environment is tempered, as is the extent to which U.S. firms that abide by environmental regulation compete with those who do not. The second part of the story, then, allows us to determine the form that attempted internationalization will take.

The success of these attempts at internationalizing U.S. domestic environmental regulations comes in part from these earlier factors, as well as from characteristics of the target states. A target state is most likely to accept regulations pushed by another state when the sending state has the ability to harm its interests if it does not adopt the regulation in question. Because of how the threat is formed, the market power the United States has over the target state in the commodity in question is an important predictor of whether the target state will agree to adopt the regulation. If a target state suffers the loss of an important market for its goods by refusing to accept a regulation, it is likely to undertake protection of the resource in question, so long as that protection is not more costly than the loss of the market would have been. But the success of market power hinges also on the belief that it will be used—threatened sanctions often are not imposed, and imposed sanctions may be removed. So the credibility of the threat comes from the continued existence of the coalition that not only pushed for the threat in the first place, but also actually benefits from its imposition. Although it might not seem that the extent of a particular Baptist-and-bootlegger coalition would vary at different stages, internationalizing legislation can indeed sometimes be used on issues that are not the ones for which it was passed.[2] Because of this use of legislation by coalitions different than those that worked to pass it, there can be variation on the credibility of the threat. This variation in credibility, in turn, contributes to variation in success of internationalization. Because we already know the form that threats will take, and because it is not difficult to observe the potential U.S. actors behind an internationalization attempt, this third part of the story allows us to predict the success of any given internationalization attempt.[3]

It should be pointed out that threats are not the only way in which the United States attempts to internationalize its domestic environmental regulations, though they do play a surprisingly large role in the issue areas examined here, in light of the general perception that environmental issues are low politics concerns and that everyone gains from environmental protection. International negotiations, sometimes of multilateral agreements, are

---

[2] If the original legislation is written broadly enough, it can be used or expanded by one aspect of the coalition to apply to other, similar, issues on which there is no coalition partner. Or, the regulation may become stricter over time, putting the industry actors out of business.

[3] Including, perhaps, one that has not yet taken place.

also used in internationalization attempts and sometimes quite successfully. Of particular interest is that the two strategies often go hand in hand. States consider or threaten unilateral economic sanctions should multilateral diplomacy fail, and sometimes, unilateral internationalization efforts result in multilateral agreements. The several recent international agreements for dolphin protection are one instance of the latter; the consideration of U.S. economic restrictions on imports of ozone-depleting substances should the negotiations at Montreal fail is an example of the former. Two lessons should be taken from this observation. First, environmental politics is still a realm of conflict, power, and threat, even if there are positive sum games to be played by states in regulating. Second, international cooperation and unilateral threats are not different realms of politics. They may not only be compatible but sometimes reinforcing.

Threats help create international environmental agreements, and help to ensure full participation in, implementation of, and compliance with those agreements. Unilateral action may be taken in support of multilateral negotiation, as seen in several of the internationalization cases here. Often the domestic legislation authorizing economic sanctions calls for multilateral negotiations, as did the sea-turtle-protection sanctions and the measures to restrict imports of fish caught by driftnets. Both of these measures have resulted in multilateral agreements to address the environmental problem in question, as have other of the threats examined here. In other issue areas as well, the United States has used economic threats or restrictions in support of multilateral action. Section 301, though controversial, was at least presented as a tool for the support of international free trade, and some actions taken pursuant to it did result in multilateral trade agreements. Some argue, for instance, that the conclusion of an agreement within the Uruguay Round trade negotiations on agriculture was made possible by U.S. threats of trade restrictions on European Community exports.[4] Multilateral agreements may result from unilateral action even when they were not the initial goal, as targets determine that they would rather play a role in their own regulation. Some of the interim agreements negotiated in the wake of the U.S. demands for access to tuna within EEZs of other states have this character. U.S. unilateral threats can result in multilateral action.

---

[4] Charles Iceland, "European Union: Oilseeds," in *Reciprocity and Retaliation in U.S. Trade Policy,* ed. Thomas O. Bayard and Kimberly Ann Elliott (Washington, D.C.: Institute for International Economics, 1994), p. 209.

That success in internationalization can be attributed largely to the relative market power of the states involved (combined with a credible threat of using it) leaves some questions unanswered. In particular, the internationalizing state has choices about what types of threats it makes in attempts to convince other states to adopt environmental regulations—why would it ever make threats on the basis of market power insufficient to succeed?

This question takes us back to the coalition at Stage I and the reasons why each coalition member works for internationalization. The form of the threat is determined by the type of environmental regulation in question and the coalition that supports its internationalization. The suggestion of an alternate form of threat, even if such a threat would rely on greater market power, would be unlikely to meet the approval of the coalition that attempted internationalization in the first place. And at the threat stage, such a threat might be less credible to the target state, since the coalition in support of internationalization would not benefit from the imposition of the threat. Also, the target states vary in the extent to which they rely on the United States as a market for the good in question. There is almost nothing the United States can do by way of threat to convince French Guiana to adopt sea turtle protection, since it does not sell any of its shrimp in the United States, but other states that do are vulnerable. The extent to which a state relies on U.S. markets will determine the ability of the U.S. to threaten it successfully with these types of measures.

What this all means is that success of internationalization attempts will vary. But because we know the incentives behind internationalization, it is possible to predict which domestic U.S. regulations are likely candidates for attempts at internationalization, and which would likely be adopted internationally should the United States try to push them forward internationally.

## Lessons for Environmentalists

This study suggests that the environmentalist slogan "Think Globally, Act Locally" is an accurate one for those who are concerned with international environmental protection. In the stage before the ones examined in this project, environmentalists and industry clashed at the domestic level of environmental regulations. The regulations considered here as possible cases of internationalization are those that environmental activists were able to pass domestically, often over the objection of industry actors. But once these regulations passed, some industry actors gained a new incentive, due to the regulation, to join the call of environmentalists for international regulation of the environmental issue. There is some advantage, then, if one wants global environmental protection, to working to pass domestic regulation on that issue first.

A possible corollary may be that states are less likely to be environmental leaders internationally when they have not yet passed domestic regulations to address an environmental problem. The United States took on a leadership role in multilateral efforts to address ozone depletion and endangered species protection when it already had relevant regulations domestically; it has been unwilling to support international action on such issues as climate change and biodiversity protection where it does not. Although this relationship remains to be investigated more fully, this anecdotal evidence may provide additional reasons to begin regulation of international environmental issues domestically within influential states.

The type of regulation an environmentalist should work for is suggested as well by this study. Regulation of domestic industries will lead to a call for internationalization only if those industries compete internationally and if the environmental issue in question is a transboundary one. The very legislation that inspires internationalization may hamper its effectiveness if not planned carefully. Since the continued existence of industry actors is important for the influence they give at both stages of internationalization, there is some advantage to regulating, but not putting out of business, the industry actors causing environmental damage. The public policy implications of this observation are counterintuitive—in order to work successfully for regulation on the international level, it may be advantageous in some cases not to work for a complete cessation of the activity in question. In the case of whaling, for instance, the cessation of commercial whaling in the United States under the Marine Mammal Protection Act meant that there was no industry partner when pushing regulations internationally. That lack ultimately made threats less credible. Although there could conceivably be instances in which an activity is sufficiently harmful or repugnant that putting the industry actor out of business is the only acceptable alternative, it may be strategically more helpful in the long run to avoid doing so initially.

More importantly, it is useful to remember that the relationship between environmentalists and industry actors is not a given. The industry partner in this coalition does not exist before the regulation is initially created, so the way domestic regulations are targeted will have implications for their successful internationalization. If ozone-depleting substances had been immediately banned within the United States, there would have been no coalition partner for environmentalists concerned about international regulation. If the United States had decided to pursue sea turtle protection by committing to the protection of beaches where turtles lay their eggs rather than by regulating shrimp fishers, the resulting politics of internationalization would have taken on an entirely different character. In this sense, Baptists create bootleggers and should be advised to consider what types of concerned actors will contribute most in the long run to international environmental regulation.

There are also numerous ways to pursue initial domestic regulation, some of which need not begin with governmental action. Support for ozone layer protection began with U.S. domestic consumer action. Even before legislation outlawing the use of CFCs in nonessential aerosols, U.S. use of spray cans decreased by nearly two-thirds simply as a result of consumer demand.[5] Local municipalities began debating resolutions to disallow use of ozone-depleting substances. Likewise, demand for the protection of dolphins involved early consumer demand for tuna that was caught in ways that did not harm dolphins. These types of consumer actions can precede, or even potentially supplant, governmental action to regulate domestically. Even voluntary industry standards may have similar effects, even if taken in an effort to avoid stricter regulation, or to relatively disadvantage domestic competitors. Any of these avenues may be routes to internationalization.

Environmentalists should also be reminded that the type of internationalization examined here cannot be the only type pursued by those who are concerned about the protection of the international environment. If French Guiana is not vulnerable to U.S. threats as a means to change its environmental policies, other methods will have to be pursued if such a change is desired. The ozone example is again illustrative here. Although some states adopted ozone protection policies because of a feared loss of market access and others did so because they succeeded in getting their costs of compliance

5 Richard Elliott Benedick, *Ozone Diplomacy: New Directions in Safeguarding the Planet* (Cambridge, Mass.: Harvard University Press, 1991), pp. 27–28.

compensated, a number of states agreed to protect the ozone layer because they were convinced of the severity of the environmental problem. In some instances increasing scientific certainty or a demonstration of substitute processes may be useful; in other instances a change in values may be necessary. Some of these changes can take place within the type of internationalization effort described here, but there are many other parallel processes that environmental activists can undertake in an effort to support international environmental protection.

In short, the process examined here is not the only way to pursue international environmental regulation, but it is a potentially influential one. Environmentalists can take from this study a set of suggestions for how to work for domestic regulation that can later be internationalized, and for prioritization of strategies and issues for doing so.

### Lessons for Scholars of Sanctions

This study has implications as well for those who examine threats and sanctions as tools of foreign policy. The first relates to the question of the "success" of sanctions in general. This question has plagued debates about the use of sanctions, with influential theorists pronouncing that sanctions don't "work."[6] Even Gary Hufbauer, Jeffrey Schott, and Kimberly Ann Elliott, who began with a goal of rehabilitating the idea of sanctions, ultimately conclude that "although it is not true that sanctions 'never work,' they are of limited utility in achieving foreign policy goals that depend on compelling the target country to take actions it resists."[7]

The typical response to this pessimistic view of the success of sanctions is to redefine the question of what it means for sanctions to "work": to argue that often sanctions are imposed "primarily for domestic or other rhetorical purposes."[8] In these cases sanctions may be successful at mollifying a

---

[6] Some, such as Johan Galtung, even argue that they can be counterproductive. See Johan Galtung, "On the Effects of International Economic Sanctions," *World Politics* 19, no. 3 (1967): 378–416. Most who refer to the success of sanctions refer to the ability of sanctions to change the behavior of the target states.

[7] Gary Clyde Hufbauer, Jeffrey J. Schott, and Kimberly Ann Elliott, *Economic Sanctions Reconsidered: History and Current Policy*, 2nd ed. (Washington, D.C.: Institute for International Economics, 1990), p. 92.

[8] Hufbauer, Schott, and Elliott, *Economic Sanctions Reconsidered*; see also Michael P. Malloy, *Economic Sanctions and U.S. Trade* (Boston: Little, Brown, 1990).

domestic community or providing symbolic action, even when they do not do much to change the target state's policies; occasionally that is all they are intended to accomplish.

This study provides examples of a number of sets of sanctions that do "work" (both in the domestic sense and in the sense of achieving a foreign policy goal) and some suggestions about conditions under which sanctions are likely to be successful. One of the explanations for the failure of sanctions in general is that they harm the sending state and so, therefore, are not credible threats. In this study the explanation for success that relies on the credibility of threat derives from the members of the sending state actually *gaining* by the imposition of sanctions. Target states know that if they do not take the requested action, sanctions will indeed be imposed. And this study shows that economic sanctions can be a much more refined and specific tool than is often imagined. Sanctions may fail at disabling an economy and ousting a tyrannical leader, but well-specified sanctions may be able to change the way industries in target states do business, and in that way have profound effects within and beyond the borders of that country.

This study also contributes to the discussion of how international relations scholars think about power. The size of a state's GNP or military is not going to influence the willingness of a small Latin American state to increase its protection of sea turtles, especially given the improbability of going to war over endangered species. A further refinement in general is necessary when examining power over whom to do what. In this case, market power over a specific resource relating to an environmental problem is the important resource. Without it even a state as powerful as the United States cannot always get its way. Considering which aspects of power are likely to be important in which situations will be essential to understanding international influence.

The role of the world trading system is also worth considering in any discussion of economic sanctions and their effectiveness. As evidenced by the cases examined here, the disapproval of the GATT/WTO trading system does not seriously hinder the effectiveness of internationalization efforts. These processes of internationalization through trade restrictions have meanwhile taken place in a broader context of discussions about the intersection between free trade and environmental protection, as well as of a changing process of handling disputes on these issues. Efforts are underway to reform the world trading system in a way that makes it more compatible with environmental protection, at the same time that the GATT/WTO dispute resolution process has been strengthened to make it less forgiving of states that try to circumvent free trade. Developments on both these fronts are likely to have an impact upon the future use of unilateral trade restrictions in support of international environmental policy.

The response of the GATT and WTO dispute resolution processes to the cases examined here, however, indicates that the international trading system is far from treating free trade and international environmental protection through trade restrictions as inherently incompatible. In all of the cases where U.S. internationalization efforts were taken before international trade dispute settlement bodies, U.S. sanctions were found to be at least partly illegal. But that was due largely to the way they were implemented, rather than to the inherent unacceptability of trade restrictions under all circumstances. The main finding against the sea turtle sanctions, for instance, was that they were applied to a set of states for which they were not originally intended. These later target states were discriminated against because they were not given the same time frame within which to phase in sea turtle protection measures that the original target states were. The goal of the legislation, protection of sea turtles, was deemed acceptable, especially since these species were listed on the appendices of CITES.[9] Even the extraterritoriality of the U.S. actions was not inherently problematic. The second tuna/dolphin panel ruling explicitly indicated that states could undertake measures to influence environmental conservation outside national jurisdiction.[10] All panel rulings indicated the importance of multilateral negotiation as a precursor to any unilateral action but kept open the possibility that additional measures might be permissible.

Across the panel rulings there is a clear trend toward delineating the conditions under which sanctions might be accepted internationally: probably, if they are multilateral; almost certainly, if they are done in the context of an international agreement; possibly, if they are done unilaterally but after serious effort at multilateral negotiation. The dispute settlement reports do

[9] World Trade Organization, "United States—Import Prohibition of Certain Shrimp and Shrimp Products." Interim Panel Report, unpublished copy obtained from the Office of the United States Trade Representative, 1998.

[10] General Agreement on Tariffs and Trade, "United States—Restrictions on Imports of Tuna." Report of the Panel DS29/R, June 1994, pp. 891–93.

not embrace or even accept the concept of unilateral trade restrictions, but neither do they forbid such restrictions under all circumstances. Despite fears of environmental activists that the WTO is unfavorably disposed toward environmental concerns, there appears to be increasing acceptance of the idea of trade restrictions as part of a process of international environmental protection. The U.S. experience with these measures, both in the domestic favor they enjoy and in the internationalization success they engender, indicates that such measures are unlikely to disappear as a means of internationalization of domestic environmental regulations.

### Broader Applicability

To what extent does a study about U.S. internationalization of domestic environmental policies on three specific issue areas contribute anything further than explaining why whaling sanctions are unlikely to succeed or why French Guiana was able to avoid adopting U.S.-dictated sea turtle protections? There is reason to assume that the findings of this study are of relevance to thinking about how other states address international environmental policy, and to the relationship between economic and moral aspects of policymaking on a number of issues.

Although less frequent, there are internationalization attempts on environmental issues by states other than the United States. European acid rain and Canadian fishery policies have been conducted to some extent through internationalization. To what extent are these internationalization attempts similar to the ones examined here? There is anecdotal evidence that the same types of coalitions that exist in the United States to push internationalization of environmental policies also exist in other states. The examples given in chapter 1 suggest that air pollution regulations in Europe may have been pushed in part by industry actors who gained from regulation either because they already bore regulations or because they could export control technology. The political dynamics of different states may involve different processes for pushing regulation on the international level, and it would be worthwhile to look in depth at this process in other states.

A cursory glance at the recent Canada-Spain conflict over regulation of turbot suggests that a Baptist-and-bootlegger coalition (environmentalists concerned about severely declining fish resources joining Canadian fishers concerned that Spanish fishers would take all their turbot and put them out of business) was behind Canada's attempt to convince Spain to adopt the same level of turbot protection Canada already had. Clearly, not all states will be capable of successfully bringing domestic issues onto the international agenda, and not all that do will be propelled by Baptist-and-bootlegger coalitions. But the phenomenon certainly extends beyond the arena of U.S. environmental regulation.

This type of coalition may form across borders as well, in ways that can have impact upon international environmental policy. Examples abound of environmental or health activists in one country joining forces with industry actors in another. For example, efforts to restrict exports of U.S. cigarettes were led by U.S. health activists and Asian cigarette manufacturers.[11] The same phenomenon may happen within regions of a country or across states within a supranational organization like the European Union. Although the coalition may be international in scope, the industry partner will almost certainly be found within the state considering internationalizing a regulation. Environmentalists have interests that are inherently international. They want to increase the overall level of environmental protection. They can therefore be drawn from anywhere. The interests of industry actors are national. They want to improve their position vis-à-vis international competitors. They provide the domestic push for internationalizing a regulation. Although it might be more likely that they can successfully internationalize with environmental activists on their side domestically, evidence suggests that it is also possible to do so in coalition with environmentalists from elsewhere.

The next question, in turn, is the applicability of this study to issue areas other than the environment. Certainly, the moral component is clearer on environmental protection than on other issues. A likely area to which the results of this study might be extended, however, is regulatory policies in general. Regulation of working conditions or of safety of materials used on the domestic level is likely to enjoy both a moral and a competitiveness impetus for internationalization. The campaign by international labor unions to increase labor standards on ships flying flags of convenience, supported by states that require high labor standards of their own vessels, may

---

[11] David Vogel, *Trading Up: Consumer and Environmental Regulation in a Global Economy* (Cambridge, Mass.: Harvard University Press, 1995), pp. 198–200.

be one example of this phenomenon. It will be interesting to discover in which contexts such a coalition is likely to form to work for internationalization of regulation in general. It would also be useful to determine the extent to which international regulations begin with domestic regulations by one state. If domestic action often precedes international regulation, then this formulation might allow us to predict regulations that are likely to appear internationally. A brief survey of multilateral agreements in the issue areas studied here shows that often, the international regulations pursued in these issue areas began on the domestic level in one or more states.[12] Further analysis is warranted.

From an analytical perspective, this project has contributed one piece to the puzzle of understanding the sources of international regulation. From an environmental perspective it has contributed a suggestion for how to use that piece to make the process of international environmental regulation less puzzling and more effective.

---

[12] That is less true of early fisheries agreements, but seems to hold for later agreements. More analysis is needed of the roots of international or cross-national regulation in general.

# Appendix A: Potential Impact of Sea Turtle Sanctions

| State/ Year[a] | Total Exports ($US thousands) | Shrimp[b] Exports ($US thousands) | Shrimp Exports to U.S. ($US thousands) | Shrimp to U.S./ Shrimp Exports | Shrimp Exports/ Total Exports | Shrimp to U.S./ Total Exports |
|---|---|---|---|---|---|---|
| Belize | | | | | | |
| 1990 | 104,548 | *** | *** | *** | *** | *** |
| 1991 | *** | *** | *** | *** | *** | *** |
| 1992 | 110,980 | 6,743 | 4,620 | 68.5% | 6.07% | 4.16% |
| 1993 | 108,985 | 7,584 | 6,188 | 81.6% | 6.96% | 5.68% |
| 1994 | 119,395 | 13,067 | 9,807 | 75.1% | 10.94% | 8.21% |
| 1995 | 142,920 | 15,504 | 11,878 | 76.6% | 10.85% | 8.31% |
| 1996 | 153,551 | 12,061 | 10,401 | 86.2% | 7.85% | 6.77% |
| 1997 | 158,937 | 17,646 | 16,644 | 94.3% | 11.10% | 10.47% |
| Brazil | | | | | | |
| 1990 | 31,411,641 | 110,737 | 70,378 | 63.6% | 0.35% | 0.22% |
| 1991 | 31,621,751 | 118,862 | 72,119 | 60.7% | 0.38% | 0.23% |
| 1992 | 35,975,460 | 117,373 | 74,089 | 63.7% | 0.33% | 0.21% |
| 1993 | 38,700,877 | 122,526 | 79,607 | 65.0% | 0.32% | 0.21% |
| 1994 | 43,557,977 | 124,305 | 81,189 | 65.3% | 0.29% | 0.19% |
| 1995 | 46,505,382 | 111,349 | 71,843 | 64.5% | 0.24% | 0.15% |
| 1996 | 44,746,546 | 89,188 | 46,420 | 52.0% | 0.20% | 0.10% |
| China | | | | | | |
| 1995 | 148,779,565 | 1,024,766 | 158,649 | 15.5% | 0.69% | 0.11% |
| 1996 | 151,047,526 | 781,830 | 177,166 | 22.7% | 0.52% | 0.12% |

| State/<br>Year[a] | Total<br>Exports<br>($US thou-<br>sands) | Shrimp[b]<br>Exports<br>($US<br>thou-<br>sands) | Shrimp<br>Exports<br>to U.S.<br>($US<br>thou-<br>sands) | Shrimp<br>to U.S./<br>Shrimp<br>Exports | Shrimp<br>Exports/<br>Total<br>Exports | Shrimp<br>to U.S./<br>Total<br>Exports |
|---|---|---|---|---|---|---|
| Colombia | | | | | | |
| 1990 | 6,765,037 | 79,508 | 43,609 | 54.8% | 1.18% | 0.64% |
| 1991 | 7,268,642 | 96,203 | 51,861 | 53.9% | 1.32% | 0.71% |
| 1992 | 6,916,051 | 76,670 | 35,846 | 46.8% | 1.11% | 0.52% |
| 1993 | 7,454,865 | 83,629 | 45,775 | 54.7% | 1.12% | 0.61% |
| 1994 | 8,916,833 | 125,352 | 64,980 | 51.8% | 1.41% | 0.73% |
| 1995 | 10,327,778 | 119,898 | 55,431 | 46.2% | 1.16% | 0.54% |
| 1996 | *** | *** | *** | *** | *** | *** |
| Costa Rica | | | | | | |
| 1990 | 1,455,640 | 10,399 | 5,227 | 50.3% | 0.71% | 0.36% |
| 1991 | 1,627,571 | 16,919 | 8,254 | 48.8% | 1.04% | 0.51% |
| 1992 | 1,833,713 | 53,979 | 11,922 | 22.1% | 2.94% | 0.65% |
| 1993 | 1,941,721 | 48,568 | 24,931 | 51.3% | 2.50% | 1.28% |
| 1994 | 2,220,441 | 30,967 | 19,363 | 62.5% | 1.39% | 0.87% |
| 1995 | 2,701,759 | 32,955 | 13,121 | 39.8% | 1.22% | 0.49% |
| 1996 | 2,779,794 | 63,612 | 22,844 | 35.9% | 2.29% | 0.82% |
| Ecuador | | | | | | |
| 1995 | 4,361,500 | 678,649 | 417,749 | 61.6% | 15.6% | 9.58% |
| 1996 | 4,889,834 | 631,390 | 355,452 | 56.3% | 12.9% | 7.27% |
| 1997 | 5,214,143 | 874,432 | 540,728 | 61.8% | 16.8% | 10.4% |
| El Salvador | | | | | | |
| 1995 | 985,202 | 27,415 | 25,839 | 94.3% | 2.78% | 2.62% |
| 1996 | 1,024,266 | 40,833 | 38,688 | 94.7% | 3.99% | 3.78% |
| French Guiana | | | | | | |
| 1990 | 89,743 | 38,110 | 2,206 | 5.79% | 42.5% | 2.46% |
| 1991 | 69,974 | 33,684 | 0 | 0.00% | 48.1% | 0.00% |
| 1992 | 102,109 | 38,074 | 6 | 0.01% | 37.3% | 0.01% |
| 1993 | 102,139 | 26,111 | 0 | 0.00% | 25.6% | 0.00% |
| 1994 | 148,861 | 31,178 | 0 | 0.00% | 20.9% | 0.00% |
| 1995 | 158,168 | 34,887 | 0 | 0.00% | 22.1% | 0.00% |
| 1996 | *** | *** | *** | *** | *** | *** |

| State/<br>Year[a] | Total<br>Exports<br>($US thou-<br>sands) | Shrimp[b]<br>Exports<br>($US<br>thou-<br>sands) | Shrimp<br>Exports<br>to U.S.<br>($US<br>thou-<br>sands) | Shrimp<br>to U.S./<br>Shrimp<br>Exports | Shrimp<br>Exports/<br>Total<br>Exports | Shrimp<br>to U.S./<br>Total<br>Exports |
|---|---|---|---|---|---|---|
| Guatemala | | | | | | |
| 1990 | 1,162,970 | 13,517 | 12,758 | 94.4% | 1.16% | 1.10% |
| 1991 | 1,202,193 | 16,835 | 13,598 | 80.8% | 1.40% | 1.13% |
| 1992 | 1,295,291 | 19,422 | 13,868 | 71.4% | 1.50% | 1.07% |
| 1993 | 1,338,222 | 25,834 | 14,910 | 57.7% | 1.93% | 1.11% |
| 1994 | 1,502,445 | 29,833 | 21,036 | 70.5% | 1.99% | 1.40% |
| 1995 | 1,935,516 | 22,595 | 14,719 | 65.1% | 1.17% | 0.76% |
| 1996 | 2,030,734 | 26,707 | 19,051 | 71.3% | 1.32% | 0.94% |
| Guyana[c] | | | | | | |
| 1990 | 232,000 | 23,424 | 12,152 | 51.9% | 10.10% | 5.24% |
| 1991 | 292,000 | 19,875 | 16,309 | 82.1% | 6.81% | 5.58% |
| 1992 | 363,000 | 13,024 | 13,564 | 104.1% | 3.59% | 3.74% |
| 1993 | 434,000 | 11,569 | 14,497 | 125.3% | 2.66% | 3.34% |
| 1994 | 483,000 | 13,048 | 16,857 | 129.2% | 2.70% | 3.49% |
| 1995 | 502,000 | 15,249 | 18,099 | 118.7% | 3.04% | 3.61% |
| 1996 | 572,000 | 12,641 | 22,219 | 175.8% | 2.21% | 3.88% |
| Honduras | | | | | | |
| 1990 | *** | *** | *** | *** | *** | *** |
| 1991 | 616,652 | 45,859 | 45,491 | 99.2% | 7.44% | 7.38% |
| 1992 | 515,670 | 16,545 | 16,439 | 99.4% | 3.21% | 3.19% |
| 1993 | 664,759 | 54,830 | 49,853 | 90.9% | 8.25% | 7.50% |
| 1994 | 614,049 | 73,191 | 63,886 | 87.3% | 11.9% | 10.4% |
| 1995 | 656,012 | 35,807 | 32,571 | 90.1% | 5.46% | 4.97% |
| 1996 | 845,988 | 57,473 | 50,805 | 88.4% | 6.79% | 6.01% |
| Indonesia | | | | | | |
| 1995 | 45,417,982 | 1,080,776 | 52,531 | 4.86% | 2.38% | 0.16% |
| 1996 | 49,814,716 | 1,063,599 | 107,819 | 10.1% | 2.14% | 0.22% |
| Mexico | | | | | | |
| 1990 | 26,344,680 | 241,065 | 226,539 | 94.0% | 0.92% | 0.86% |
| 1991 | 28,956,722 | 257,513 | 243,767 | 94.7% | 0.89% | 0.84% |
| 1992 | 46,194,885 | 239,727 | 233,713 | 97.5% | 0.52% | 0.51% |
| 1993 | 51,886,415 | 325,711 | 322,046 | 98.9% | 0.63% | 0.62% |
| 1994 | 60,618,583 | 376,293 | 362,215 | 96.3% | 0.62% | 0.60% |
| 1995 | 79,540,655 | 537,218 | 508,610 | 94.7% | 0.68% | 0.64% |
| 1996 | 95,661,198 | 574,433 | 488,499 | 85.0% | 0.60% | 0.51% |

| State/<br>Year[a] | Total<br>Exports<br>($US thou-<br>sands) | Shrimp[b]<br>Exports<br>($US<br>thou-<br>sands) | Shrimp<br>Exports<br>to U.S.<br>($US<br>thou-<br>sands) | Shrimp<br>to U.S./<br>Shrimp<br>Exports | Shrimp<br>Exports/<br>Total<br>Exports | Shrimp<br>to U.S./<br>Total<br>Exports |
|---|---|---|---|---|---|---|
| Nicaragua | | | | | | |
| 1990 | 340,034 | 8,838 | 2,038 | 23.1% | 2.60% | 0.60% |
| 1991 | 265,785 | 13,113 | 10,237 | 78.1% | 4.93% | 3.85% |
| 1992 | 236,493 | 15,595 | 10,434 | 66.9% | 6.59% | 4.41% |
| 1993 | 267,493 | 22,377 | 16,585 | 74.1% | 8.37% | 6.20% |
| 1994 | 351,108 | 43,106 | 34,406 | 79.8% | 12.28% | 9.80% |
| 1995 | 509,212 | 69,743 | 56,574 | 81.1% | 13.69% | 11.11% |
| 1996 | 660,183 | 63,962 | 52,485 | 82.1% | 9.69% | 7.95% |
| Nigeria[d] | | | | | | |
| 1995 | 11,744,000 | 50,830 | 2,545 | 5.0% | 0.01% | 0.00% |
| 1996 | 14,836,000 | 50,830 | 610 | 1.2% | 0.00% | 0.00% |
| Panama | | | | | | |
| 1990 | 340,808 | 50,080 | 43,236 | 86.3% | 14.7% | 12.7% |
| 1991 | 452,094 | 56,178 | 48,459 | 86.3% | 12.4% | 10.7% |
| 1992 | 341,831 | 56,178 | 48,459 | 86.3% | 16.4% | 14.2% |
| 1993 | 480,912 | 62,635 | 52,000 | 83.0% | 13.0% | 10.8% |
| 1994 | 506,828 | 65,885 | 53,713 | 81.5% | 13.0% | 10.6% |
| 1995 | 539,823 | 81,389 | 67,401 | 82.8% | 15.1% | 12.5% |
| 1996 | *** | *** | *** | *** | *** | *** |
| Suriname | | | | | | |
| 1990 | 472,769 | 35,969 | 4 | 0.01% | 7.61% | 0.00% |
| 1991 | 359,247 | 32,173 | 8 | 0.02% | 8.96% | 0.00% |
| 1992 | 357,147 | 23,117 | 7 | 0.03% | 6.47% | 0.00% |
| 1993 | *** | *** | *** | *** | *** | *** |
| 1994 | 520,451 | 45,591 | 4 | 0.01% | 8.76% | 0.00% |
| 1995 | 475,429 | 37,086 | 45 | 0.12% | 7.80% | 0.01% |
| 1996 | *** | *** | *** | *** | *** | *** |
| Thailand | | | | | | |
| 1995 | 56,344,556 | 2,411,552 | 559,543 | 23.2% | 4.28% | 0.99% |
| 1996 | *** | *** | *** | *** | *** | *** |
| 1997 | 58,085,783 | 1,885,157 | 484,265 | 25.7% | 3.24% | 0.83% |

| State/<br>Year[a] | Total<br>Exports<br>($US thou-<br>sands) | Shrimp[b]<br>Exports<br>($US<br>thou-<br>sands) | Shrimp<br>Exports<br>to U.S.<br>($US<br>thou-<br>sands) | Shrimp<br>to U.S./<br>Shrimp<br>Exports | Shrimp<br>Exports/<br>Total<br>Exports | Shrimp<br>to U.S./<br>Total<br>Exports |
|---|---|---|---|---|---|---|
| Trinidad & Tobago | | | | | | |
| 1990 | 1,984,993 | 1,386 | 752 | 54.3% | 0.07% | 0.04% |
| 1991 | 1,984,994 | 1,367 | 753 | 55.1% | 0.07% | 0.04% |
| 1992 | 1,868,943 | 1,514 | 1,002 | 66.2% | 0.08% | 0.05% |
| 1993 | 1,662,107 | 2,047 | 1,424 | 69.6% | 0.12% | 0.09% |
| 1994 | 1,954,282 | *** | *** | *** | *** | *** |
| 1995 | 2,466,984 | *** | *** | *** | *** | *** |
| 1996 | 2,456,439 | 2,391 | 507 | 21.2% | 0.10% | 0.02% |
| Venezuela | | | | | | |
| 1990 | 18,044,254 | 37,864 | 21,861 | 57.7% | 0.21% | 0.12% |
| 1991 | 15,129,876 | 48,139 | 26,486 | 55.0% | 0.32% | 0.18% |
| 1992 | 14,235,322 | 36,013 | 16,849 | 46.8% | 0.25% | 0.12% |
| 1993 | 15,208,136 | 40,789 | 30,515 | 74.8% | 0.27% | 0.20% |
| 1994 | 16,649,658 | 42,163 | 33,894 | 80.4% | 0.25% | 0.20% |
| 1995 | 18,914,219 | 31,246 | 23,808 | 76.2% | 0.17% | 0.13% |
| 1996 | 22,674,450 | 39,850 | 29,174 | 73.2% | 0.18% | 0.13% |
| 1997 | 21,657,667 | 76,430 | 63,660 | 83.3% | 0.35% | 0.29% |

*Source:* United Nations *Commodity Trade Statistics* for the years in question.

[a] The year considered is the one *before* sanctions were threatened or imposed, so that the measurement does not reflect the possible sanctions.

[b] As a proxy for shrimp exports, exports of shellfish (SITC category 036) is used. This proxy is not perfect, but for most of these states, shrimp is their main shellfish export. Examination of shrimp export information for those from whom it is available (not used here because there is not comparable data across all cases) indicates that it is an acceptable proxy. More problematic is that this measure does not distinguish between farmed shrimp and caught shrimp, which the export restrictions do. (The TED requirement initially only applies to fish caught from shrimping boats in the certain regions, and the export restrictions for not adopting standards equivalent to those of the United States apply only to caught shrimp.) It will be noted separately which states have especially important shrimp aquaculture industries.

[c] The information on total shrimp exports for this country comes from the "crustaceans" category on the FAO database (http://www.apps.fao.org, date visited: 21 April 1999). The information on shrimp exports to the United States comes from the "shrimp imports" category on the National Marine Fisheries Service Database (http://www.nmfs.gov, date visited: 21 April 1999). That explains data discrepancies between the categories. The magnitude is nevertheless indicative.

d The information on total shrimp exports for this country comes from the "crustaceans" category on the FAO database (http://www.apps.fao.org, date visited: 21 April 1999). The information on shrimp exports to the United States comes from the "shrimp imports" category on the National Marine Fisheries Service Database (http://www.nmfs.gov, date visited: 21 April 1999). That explains data discrepancies between the categories. The magnitude is nevertheless indicative.

*** Data not available, either because that state does not report trade information to the United Nations in general or because the United Nations has not compiled that particular information for that year.

# Appendix B: Potential Impact of Whaling Sanctions

| State/Year[a] | Total Exports ($US thousands) | Fish Exports ($US thousands) | Fish Exports to U.S. ($US thousands) | Fish to U.S./ Fish Exports | Fish Exports/ Total Exports | Fish to U.S./ Total Exports |
|---|---|---|---|---|---|---|
| Canada |  |  |  |  |  |  |
| 1995[b] | 165,836,776 | 2,151,048 | 1,226,317 | 57.0% | 1.30% | 0.74% |
| Chile |  |  |  |  |  |  |
| 1977 | 2,138,421 | 34,719 | 10,512 | 30.3% | 1.62% | 0.49% |
| Iceland |  |  |  |  |  |  |
| 1985 | 813,832 | 533,167 | 202,954 | 38.1% | 65.51% | 24.96% |
| 1986 | 1,550,000 | 769,943 | 216,170 | 28.1% | 49.67% | 13.95% |
| 1987 | 1,375,113 | 993,636 | 222,397 | 22.4% | 72.26% | 16.20% |
| Japan |  |  |  |  |  |  |
| 1973 | 36,931,398 | 529,699 | 221,230 | 41.7% | 1.43% | 0.60% |
| 1983 | 146,927,471 | 663,400 | 197,456 | 29.8% | 0.45% | 0.13% |
| 1987 | 229,221,230 | 732,180 | 228,603 | 31.2% | 0.32% | 0.10% |
| 1994 | 395,599,979 | 717,993 | 133,291 | 18.6% | 0.18% | 0.04% |
| Korea, Rep. of |  |  |  |  |  |  |
| 1977 | 9,986,022 | 692,969 | 105,822 | 15.27% | 6.94% | 1.06% |
| 1979 | 14,951,794 | 794,960 | 99,854 | 12.6% | 5.31% | 0.67% |
| 1985 | 30,282,840 | 790,807 | 110,733 | 14.0% | 2.61% | 0.37% |
| Norway |  |  |  |  |  |  |
| 1985 | 19,941,156 | 817,191 | 130,247 | 15.94% | 4.10% | 0.65% |
| 1987 | 21,508,531 | 1,410,017 | 203,206 | 14.41% | 6.56% | 0.94% |
| 1989 | 27,029,737 | 1,521,921 | 142,542 | 9.37% | 5.63% | 0.54% |
| 1991 | 34,047,896 | 2,197,846 | 63,812 | 2.9% | 6.46% | 0.34% |
| 1992 | 35,137,086 | 2,327,537 | 62,090 | 2.7% | 6.62% | 0.77% |

| State/ Year[a] | Total Exports ($US thou- sands) | Fish Exports ($US thou- sands) | Fish Exports to U.S. ($US thou- sands) | Fish to U.S./ Fish Exports | Fish Exports/ Total Exports | Fish to U.S./ Total Exports |
|---|---|---|---|---|---|---|
| Peru | | | | | | |
| 1977 | 1,805,251[c] | 38,813 | 5,523 | 14.23% | 2.15% | 0.31% |
| Spain | | | | | | |
| 1979 | 13,102,517 | 279,710 | 18,016 | 6.44% | 2.13% | 0.14% |
| Taiwan | | | | | | |
| 1979 | *** | *** | *** | *** | *** | *** |
| USSR | | | | | | |
| 1973 | *** | *** | *** | *** | *** | *** |
| 1984 | *** | *** | *** | *** | *** | *** |

*Source:* United Nations *Commodity Trade Statistics* for the years in question.

[a] The year considered is the one *before* sanctions were threatened or imposed, so that the measurement does not reflect the possible sanctions.

[b] Data here are from 1994, the most recent year available.

[c] This number is from 1978.

*** Data not available, either because that state does not report trade information to the United Nations in general or because the United Nations has not compiled that particular information for that year.

# Bibliography

Ackerman, Bruce, and William T. Hassler. 1981. *Clean Coal/Dirty Air*. New Haven: Yale University Press.

Akehurst, Michael. 1987. *A Modern Introduction to International Law*. 6th ed. London: Harper Collins Academic.

Alagappan, Meena. 1990. "The United States' Enforcement of the Convention on International Trade in Endangered Species of Wild Fauna and Flora." *Northwestern Journal of International Law and Business* 10 (winter): 541–68.

"American Tunaboat Association . . ." 1977. *New York Times,* 1 January: 6.

Andresen, Steinar. 1989. "Science and Politics in the International Management of Whales." *Marine Policy* 13 (April): 99–117.

Arrow, Kenneth. 1963. *Social Choice and Individual Values*. 2nd ed. New York: Wiley.

Axelrod, Robert. 1984. *The Evolution of Cooperation*. New York: Basic Books.

Bachrach, Peter, and Morton Baratz. 1962. "Two Faces of Power." *American Political Science Review* 56: 947–52.

Baldwin, David A. 1985. *Economic Statecraft*. Princeton, New Jersey: Princeton University Press.

Baldwin, David A. 1989. *Paradoxes of Power*. New York: Basil Blackwell, 1989.

Barkin, J. Samuel, and Elizabeth R. DeSombre. 1997. "Unilateralism and Multilateralism in International Environmental Politics." Paper presented at the American Political Science Association Annual Meeting, August.

Barkin, J. Samuel, and George Shambaugh, eds. 1999. *Anarchy and the Environment: The International Relations of Common Pool Resources*. Albany: SUNY Press.

Barbera, Anthony J., and Virginia D. McConnell. 1990. "The Impact of Environmental Regulations on Industry Productivity." *Journal of Environmental Economics and Management* 18: 56–65.

Barnett, Richard. 1986. "Business Forum: Saving the Earth's Ozone Layer; The U.S. Can't Do the Job All Alone." *New York Times,* Section 3: 2 (Lexis/Nexis).

Beck, Simon. 1996. "US Recognises Move to End Wildlife Trade." *South China Morning Post* 13 September: 4 (Lexis/Nexis).

Ben Shaul, D'vora. 1997. "High Stakes." *Jerusalem Post,* 21 September: 7 (Lexis/Nexis).

Benedick, Richard Elliot. 1991. *Ozone Diplomacy: New Directions in Safeguarding the Planet.* Cambridge, Mass.: Harvard University Press.

Berrill, Michael. 1997. *The Plundered Seas: Can the World's Fish Be Saved?* San Francisco: Sierra Club Books.

Bhagwati, Jagdish. 1990. "Aggressive Unilateralism: An Overview." In *Aggressive Unilateralism: America's 301 Trade Policy and the World Trading System,* edited by Jagdish Bhagwati and Hugh T. Patrick, pp. 1–45. Ann Arbor: University of Michigan Press.

Bhagwati, Jagdish, and Hugh T. Patrick, eds. 1990. *Aggressive Unilateralism: America's 301 Trade Policy and the World Trading System.* Ann Arbor: University of Michigan Press.

Bienen, Henry, and Robert Gilpin. 1980. "Economic Sanctions as a Response to Terrorism." *Journal of Strategic Studies* 3 (May): 89–98.

Black, Dorothy J. 1992. "International Trade v. Environmental Protection: The Case of the U.S. Embargo on Mexican Tuna." *Law and Policy in International Business* 24 (1) (Lexis/Nexis).

Bonanno, Alessandro, and Douglas Constance. 1996. *Caught in the Net: The Global Tuna Industry, Environmentalism, and the State.* Lawrence: University Press of Kansas.

Bradsher, Keith. 1991. "Sea Turtles Put New Friction in U.S.-Japan Trade Quarrels." *New York Times,* 17 May: A1 (Lexis/Nexis).

Brooke, James. 1992. "America—Environmental Dictator?" *New York Times,* 3 May: C7 (Lexis/Nexis).

Brooke, James. 1992. "10 Nations Reach Accord on Saving Dolphins." *New York Times,* 12 May: C4.

Brown, Gardner, Jr., and Wes Henry. 1989. *The Economic Value of Elephants.* London: London Environmental Economics Center.

Bryner, Gary C. 1995. *Blue Skies, Green Politics: The Clean Air Act of 1990 and Its Implementation.* 2nd ed. Washington, D.C.: CQ Press.

Buck, Eugene H. 1990. "Turtle Excluder Devices: Sea Turtles and/or Shrimp?" *CRS Report for Congress,* 28 November.

Cairncross, Francis. 1991. *Costing the Earth: The Challenge to Governments, the Opportunities for Business.* Boston: Harvard Business School Press.

Caldwell, Lynton Keith. 1984. *International Environmental Policy: Emergence and Dimensions.* Durham, North Carolina: Duke University Press.

Caron, David. 1989. "International Sanctions, Ocean Management, and the Law of the Sea: A Study of Denial of Access to Fisheries." *Ecology Law Quarterly* 16: 311–54.

"The Center for Marine Conservation Commends the Clinton Administration." 1998. *U.S. Newswire,* 30 May (Lexis/Nexis).

Cevallos, Diego. 1997. "Fisheries: Mexico Dissatisfied with End of U.S. Tuna Embargo." *Inter Press Service,* 31 July (Lexis/Nexis).

Chayes, Abram, and Antonia Handler Chayes. 1992. "Extra-Treaty Sanctions." Presented at the Harvard/MIT Seminar on International Institutions and Political Economy, 10 December, unpublished draft.

Chayes, Abram, and Antonia Handler Chayes. 1995. *The New Sovereignty: Compliance with International Regulatory Agreements.* Cambridge, Mass.: Harvard University Press.

"Chemical Giant Raps EPA Action." 1980. *U.P.I.*, 16 April (Lexis/Nexis).

Clinton, William J. 1996. "Message to Congress on Japanese Whaling Activities." *Weekly Compilation of Presidential Documents* 32 (233) 10 February (Lexis/Nexis).

Clinton, William J. 1997. "Message to Congress on Canadian Whaling Activities." *Weekly Compilation of Presidential Documents* 33 (175) 10 February (Lexis/Nexis).

"Clinton Urged to Punish China, Taiwan over Wildlife." 1993. *Reuter Library Report,* 4 November (Lexis/Nexis).

Coleman, Bruce. 1977. "Troubled Waters." *Forbes,* 1 April: 56.

Collier, Robert. 1997. "Mexican Fishermen Relieved by Tuna Deal." *San Francisco Chronicle,* 30 July: A8 (Lexis/Nexis).

"Commerce Department Says South Korea and Taiwan Violate Driftnet Act." 1989. *U.P.I.*, 29 June (Lexis/Nexis).

Conybeare, John. 1986. "Trade Wars; A Comparative Study of Anglo-Hanse, Franco-Italian, and Hawley-Smoot Conflicts." in *Cooperation under Anarchy,* edited by Kenneth Oye,147-72. Princeton: Princeton University Press.

"Countries Agree on World's First Sea Turtle Treaty." 1996. *Marine Conservation News* 8, no. 4: 1, 16.

"Court in Papua New Guinea Seizes Fishing Ship from U.S." 1982. *New York Times,* 28 February: 16.

Crumm, Eileen M. 1995. "The Value of Economic Incentives in International Politics." *Journal of Peace Research* 32, no. 3: 313–31.

Cunningham, William P., and Barbara Woodworth Saigo. 1997. *Environmental Science: A Global Concern.* Dubuque, Iowa: Wm. C. Brown.

Curtain, Patty. 1990. "Annual Shrimping Kill: 44,000 Turtles." *St. Petersburg Times,* 18 May: 1A (Lexis/Nexis).

Dahl, Robert. 1984. *Modern Political Analysis.* 4th ed. Englewood Cliffs, N.J.: Prentice Hall.

DeKeiffer, Donald E. 1989. "Pyrrhic Victory in 'Driftnet War.'" *Journal of Commerce,* 24 July: 8A (Lexis/Nexis).

Denison, Edward P. 1979. *Accounting for Slower Economic Growth: The United States in the 1970s.* Washington, D.C.: Brookings Institution.

DeSombre, Elizabeth R. 1993. "Participation in the International Whaling Commission." Paper presented at the Program on Nonviolent Sanctions, Center for International Affairs, Harvard University. October.

DeSombre, Elizabeth R. 1995. "Baptists and Bootleggers for the Environment: The Origins of United States Unilateral Sanctions." *Journal of Environment and Development* 4 (1): 53–75.

DeSombre, Elizabeth R. 1999. "Tuna Fishing and Common Pool Resources." In *Anarchy and the Environment,* edited by J. Samuel Barkin and George Shambaugh, 51–69. Albany: SUNY Press.

DeSombre, Elizabeth R., and Joanne Kauffman. 1996. "Montreal Protocol Multilateral Fund: Partial Success Story." In *Institutions for Environmental Aid: Pitfalls and Promise,* edited by Robert O. Keohane and Marc Levy, 89–126. Cambridge, Mass.: MIT Press.

Doxey, Margaret P. 1987. *International Sanctions in Contemporary Perspective.* New York: St. Martin's Press.

Dumanski, Diane. 1990. "In Shift, U.S. to Aid World Fund on Ozone." *Boston Globe,* 16 June: 1.

Duncan, Alex. 1986. "Aid Effectiveness in Raising Adaptive Capacity in Low-Income Countries." In *Development Strategies Reconsidered,* edited by John P. Lewis and Valeriana Kallab, 129–152. New Brunswick, N.J.: Transaction Books.

Dunne, Nancy. 1999. "Legal Wrangle Engulfs U.S. Shrimp Dispute." *Financial Times,* 14 April: 5.

Durenberger, David. 1991. "Air Toxics: The Problem." *EPA Journal* (January/February): 30–31.

Earnshaw, Graham. 1979. "China Today . . ." *Reuters,* 23 September (Lexis/Nexis).

Earth Island Institute. No date. "Can We Stop Mexico's Shocking Slaughter of Endangered Sea Turtles?" Mailing, also in newspapers as ads.

Easterbrook, Gregg. 1994. "Forget PCBs. Radon. Alar." *New York Times Magazine,* 11 September: 60–63.

Easter-Pilcher, Andrea Lee. 1993. "Analysis of the Listing of Species as Endangered or Threatened under the Endangered Species Act." Ph.D. diss., Montana State University.

Ellison, Katherine. 1991. "Mexican Fleet, U.S. Groups Entangled in 'Tuna War.'" *Orange County Register,* 7 November: A32 (Lexis/Nexis).

Eltis, David, and James Walvin, eds. 1981. *The Abolition of the Atlantic Slave Trade.* Madison: University of Wisconsin Press.

"Environment: Sanctions against China and Taiwan to Save the Rhino." 1993. *Inter Press Service,* 7 September (Lexis/Nexis).

Environmental Protection Agency. 1993. *International Trade in Environmental Protection Equipment: An Assessment of Existing Data.* EPA 230-R-93-006. July.

Environmental Protection Agency. 1994. *Toxics Release Inventory.* Washington, D.C.: EPA.

Environmental Protection Agency. 1996. *The Benefits and Costs of the Clean Air Act, 1970 to 1990.* Prepared for the U.S. Congress. Draft. October.

"EPA Changes for RFG Imports." 1997. *Oil and Gas Journal* 12 May (Lexis/Nexis).

Esty, Daniel C., and Michael E. Porter. 1998. "Industrial Ecology and Competitiveness: Strategic Implications for the Firm." *Journal of Industrial Ecology* 2 (1): 35–43.

Favre, David S. 1989. *International Trade in Endangered Species: A Guide to CITES.* Dordrecht: Martinus Nijhoff Publishers.

Fish and Wildlife Service. 1996. "Fact Sheet on Threatened and Endangered Species," http://www.fws.gov/bio-salm.html; date visited: 26 June.

Fitzgerald, Sarah. 1989. *International Wildlife Trade: Whose Business Is It?* Washington, D.C.: World Wildlife Fund.

Franck, Thomas M. 1990. *The Power of Legitimacy among Nations.* New York: Oxford University Press.

Fraser, John. 1981. "The Politics of a Shared Environment." In *Clean Air for North Americans: Acid Rain in Candian-American Relations.* Report of a Conference Sponsored by the Canadian Studies Program, Columbia University, and issued jointly with The World Environment Center, New York City, June.

Friends of the Earth. 1990. *Funding Change: Developing Countries and the Montreal Protocol.* London: Friends of the Earth.

"FYI." 1981. *PR Newswire,* 17 February (Lexis/Nexis).

Galtung, Johan. 1967. "On the Effects of International Economic Sanctions: With Examples from the Case of Rhodesia." *World Politics* 19, no. 3: 378–416.

General Agreement on Tariffs and Trade. 1983. "United States—Prohibition of Tuna and Tuna Products from Canada." *Basic Instruments and Selected Documents* 91 107–9 (29th Supplement).

General Agreement on Tariffs and Trade. 1991. "Dispute Settlement Panel Report on United States Restrictions on Imports of Tuna [Submitted to the Parties 16 August 1991]." *International Legal Materials* 30: 1594–1623.

General Agreement on Tariffs and Trade. 1992. "United States: Restrictions on Imports of Tuna." Report of the Panel DS21/R, reprinted in General Agreement on Tariffs and Trade, *Basic Instruments and Selected Documents* Supplement no. 39, Protocols, Decisions, Reports 1991–92.

General Agreement on Tariffs and Trade. 1994. "United States—Restrictions on Imports of Tuna." Report of the Panel DS29/R, June.

Gibbons-Fly, Bill. 1994. Interview. United States Department of State. April 1.

Gilpin, Robert. 1975. *U.S. Power and the Multinational Corporation.* New York: Basic Books.

Gilpin, Robert. 1981. *War and Change in World Politics.* Cambridge: Cambridge University Press.

Gilpin, Robert. 1987. *The Political Economy of International Relations*. Princeton: Princeton University Press.

Glennon, Michael J. 1990. "Has International Law Failed the Elephant?" *American Journal of International Law* 84: 1–43.

Global Commons Institute. 1995. "Intergovernmental Panel on Climate Change, Working Group Three, A Second Defeat for Economists of Value on Human Life." *Press Release*, 10 October.

Golden, Arthur. 1985. "U.S. Plan to Lift Tuna Ban Told." *San Diego Union Tribune*, 6 September: A1 (Lexis/Nexis).

Golich, Vicki L., and Terry Forrest Young. 1993. "United States-Canadian Negotiations for Acid Rain Controls." Case 452, *Pew Case Studies in International Affairs*. Institute for the Study of Diplomacy, Pew Case Studies Center, Georgetown University.

Gore, Al. 1992. *Earth in the Balance: Ecology and the Human Spirit*. Boston: Houghton Mifflin.

Grant, Lauren. 1996. "Panama to Clamp Down on Tuna Fish Violators." *Reuters Financial Service*, 2 December (Lexis/Nexis).

Gray, Colin S. 1977. *The Geopolitics of the Nuclear Era: Heartland, Rimlands, and the Technical Revolution*. New York: Crane, Russak.

Gray, Colin S. 1988. *The Geopolitics of Super Power*. Lexington: University Press of Kentucky.

Gray, Wayne B. 1987. "The Cost of Regulation: OSHA, EPA, and the Productivity Slowdown." *American Economic Review* 77: 998–1006.

Haas, Peter M. 1990. *Saving the Mediterranean: The Politics of International Environmental Cooperation*. New York: Columbia University Press.

Haas, Peter M. 1992. "Banning Chlorofluorocarbons: Epistemic Community Efforts to Protect Stratospheric Ozone." *International Organization* 46, no. 1: 187–224.

Haas, Peter M. 1992. "Introduction: Epistemic Communities and International Policy Coordination." *International Organization* 46, no. 1: 1–35.

Haas, Peter M., Robert O. Keohane, and Marc A. Levy, eds. 1993. *Institutions for the Earth: Sources of Effective International Environmental Protection*. Cambridge, Mass.: MIT Press.

Haefele, Edwin T. 1973. *Representative Government and Environmental Management*. Baltimore: Johns Hopkins University Press, for Resources for the Future.

Hager, George. 1990. "The 'White House Effect' Opens a Long-Locked Political Door." *Congressional Quarterly Weekly Report*, 20 January: 139–44.

Hardin, Garrett. 1968. "The Tragedy of the Commons." *Science* 162(3859): 1243–1248.

Hays, Constance L. 1999. "Government Loosens Standards on Tuna Deemed 'Dolphin Safe.'" *New York Times*, 30 April: 18.

Helme, Ned, and Chris Neme. 1991. "Acid Rain: The Problem." *EPA Journal* (January/February): 18–20.

Hemley, Ginette, ed. 1994. *International Wildlife Trade: A CITES Sourcebook.* Washington, D.C.: Island Press.

Hogan, David. 1999. Interview. United States Department of State. 17 May.

Houck, Oliver A. 1993. "The Endangered Species Act and Its Implementation by the U.S. Departments of Interior and Commerce." *University of Colorado Law Review* 64: 277–370.

Hufbauer, Gary Clyde, Jeffrey J. Schott, and Kimberly Ann Elliott. 1990. *Economic Sanctions Reconsidered: History and Current Policy.* 2nd ed. Washington, D.C.: Institute for International Economics.

Hurrell, Andrew, and Benedict Kingsbury. 1992. *The International Politics of the Environment.* Oxford: Clarendon Press.

Hurrell, Andrew, and Benedict Kingsbury. 1992. "The International Politics of the Environment: An Introduction." In *The International Politics of the Environment,* edited by Andrew Hurrell and Benedict Kingsbury. Oxford: Clarendon Press.

Hyde, Alan. 1983. "The Concept of Legitimation in the Sociology of Law." *Wisconsin Law Review* (March/April): 379–426.

Iceland, Charles. 1994. "European Union: Oilseeds." In *Reciprocity and Retaliation in U.S. Trade Policy,* edited by Thomas O. Bayard and Kimberly Ann Elliott, 209–32. Washington, D.C.: Institute for International Economics.

Inter-American Convention for the Protection and Conservation of Sea Turtles. 1996.

Inter-American Tropical Tuna Commission. 1952–80. *IATTC Annual Reports.* La Jolla, Calif.: IATTC.

International Whaling Commission. 1993. *Verbatim Report* (1993 Annual Meeting).

Irwin, Aisling. 1999. "Ozone Layer Vanishing Faster than Predicted." *Daily Telegraph,* 30 April: 11 (Lexis/Nexis).

Jaffe, Adam B., Steven R. Peterson, Paul R. Portnoy, and Robert N. Stavins. 1993. "Environmental Regulations and International Competitiveness: What Does the Evidence Tell Us?" Draft, 21 December.

Jervis, Robert. 1984. *The Illogic of American Nuclear Strategy.* Ithaca: Cornell University Press.

Jorgenson, Dale, and Peter J. Wilcoxen. 1992. "Impact of Environmental Legislation on U.S. Economic Growth, Investment, and Capital Costs." In *U.S. Environmental Policy and Economic Growth: How Do We Fare?* edited by Donna L. Bodsky, 1–39. Washington, D.C.: American Council for Capital Formation.

Joseph, James, and Joseph W. Greenough. 1979. *International Management of Tuna, Porpoise, and Billfish: Biological, Legal, and Political Aspects.* Seattle: University of Washington Press.

Joyner, Christopher C. 1999. "Managing Common-Pool Marine Living Resources: Lessons from the Southern Ocean Experience." In *Anarchy and the Environment:*

*The International Relations of Common-Pool Resources,* edited by J. Samuel Barkin and George Shambaugh, 70–96. Albany: SUNY Press.

Kalt, Joseph P. 1988. "The Impact of Domestic Environmental Regulatory Policies on U.S. International Competitiveness." In *International Competitiveness,* edited by A. Michael Spence and Heather A. Hazard, 221–262. Cambridge, Mass.: Ballinger Publishing.

Kamieniecki, Sheldon, ed. 1993. *Environmental Politics in the International Arena: Movements, Parties, Organizations, and Policy.* Albany: State University of New York Press.

Karey, Gerald. 1996. "Clean Air Study: Big Costs, but Worth It." *Platt's Oligram News,* 11 June: 4 (Lexis/Nexis).

Kennedy, Paul. 1987. *The Rise and Fall of the Great Powers: Economic Change and Military Conflict from 1500 to 2000.* New York: Random House.

Kenworthy, Tom. 1993. "U.S. Pressures China, Taiwan on Animal Trade." *Washington Post,* 10 June: A28 (Lexis/Nexis).

Keohane, Robert O. 1984. *After Hegemony: Cooperation and Discord in the World Political Economy.* Princeton: Princeton University Press.

Keohane, Robert O., Peter M. Haas, and Marc A. Levy. 1993. "The Effectiveness of International Environmental Institutions." In *Institutions for the Earth: Sources of Effective International Environmental Protection,* edited by Peter M. Haas, Robert O. Keohane, and Marc A. Levy, 3–24. Cambridge, Mass.: MIT Press.

Keohane, Robert O., and Marc A. Levy, eds. 1996. *Institutions for Environmental Aid: Pitfalls and Promise.* Cambridge, Mass.: MIT Press.

Keohane, Robert O., and Helen V. Milner. 1996. *Internationalization and Domestic Politics.* Cambridge: Cambridge University Press.

Keohane, Robert O., and Elinor Ostrom, eds. 1995. *Local Commons and Global Interdependence.* London: Sage Publications.

Kibel, Paul Stanton. 1994. "GATT Fouls the Air." *Recorder,* 14 November: 8 (Lexis/Nexis).

Klinkenberg, Jeff. 1990. "Stubborn Shrimpers May Face Consumer Backlash over Turtles." *St. Petersburg Times,* 8 April: 5D (Lexis/Nexis).

Klotz, Audie. 1995. "Norms Reconstituting Interests." *International Organization* 49 (summer): 451–78.

Knorr, Klaus. 1977. "International Economic Leverage and Its Uses." In *Economic Issues and National Security,* edited by Klaus Knorr and Frank N. Trager, 99–126. Lawrence Kan.: Regents Press of Kansas, for the National Security Education Program.

Knorr, Klaus, and Frank N. Trager, eds. 1977. *Economic Issues and National Security.* Lawrence, Kan.: Regents Press of Kansas, for The National Security Education Program.

Kohn, Katheryn A., ed. 1991. *Balancing on the Brink of Extinction: The Endangered Species Act and Lessons for the Future.* Washington, D.C.: Island Press.

Korn, P. 1992. "The Case for Preservation." *Nation,* 30 March: 415–17.

Kosloff, Laura H., and Mark C. Trexler. 1987. "The Convention on International Trade in Endangered Species: Enforcement Theory and Practice in the United States." *Boston University International Law Journal* 5 (fall): 327–61.

Krasner, Stephen. 1976. "State Power and the Structure of International Trade." *World Politics* 28, no. 3: 317-43.

Krasner, Stephen. 1977. "Domestic Constraints on International Economic Leverage." In *Economic Issues and National Security,* edited by Klaus Knorr and Frank N. Trager, 160–181. Lawrence, Kan.: Regents Press of Kansas, for The National Security Education Program.

Kurlansky, Mark. 1984. "U.S., Mexico Talk Fishing Rights as the Tuna Market Ebbs." *San Diego Union-Tribune,* 31 May: C4 (Lexis/Nexis).

Kurlansky, Mark. 1997. *Cod: A Biography of the Fish That Changed the World.* Toronto: A. A. Knopf.

Lacey, Michael J., ed. 1991. *Government and Environmental Politics: Essays on Historical Developments since World War Two.* Washington, D.C.: Woodrow Wilson Center Press; and Baltimore: Johns Hopkins University Press.

LaFranchi, Howard. 1996. "Shrimp Lovers, Take Note." *Christian Science Monitor,* 29 April: 1 (Lexis/Nexis).

Lammi, Elmer W. 1987. "Washington News." *U.P.I.,* 30 April (Lexis/Nexis).

"LATAM-Fishing: Region Seeks to Avoid Repeat of Tuna Embargo." 1997. *Inter Press Service,* 1 September (Lexis/Nexis).

Lancaster, John. 1991. "Endangered Sea Turtle Seen Jeopardized by Japan." *Washington Post,* 19 January: A3.

Levy, Marc. 1992. "Political Science and the Question of Effectiveness of International Environmental and Resource Agreements: A Status Report." Prepared for the Workshop on International Environmental and Resource Agreements, Fritdjof Nansen Institute, 19–20 October.

Levy, Marc. 1993. "European Acid Rain: The Power of Tote-Board Diplomacy." In *Institutions for the Earth: Sources of Effective International Environmental Protection,* edited by Peter M. Haas, Robert O. Keohane, and Marc A. Levy, 75–132. Cambridge, Mass.: MIT Press.

Lippman, Thomas W. 1996. "Italy Faces Cutoff of Exports to U.S." *Washington Post,* 14 March: A24 (Lexis/Nexis).

Lipschutz, Ronnie D. 1995. "Who Knows? Local Knowledge and Global Governance." Paper presented at the 1995 Annual Meeting of the International Studies Association, Chicago, Ill., 21–25 February.

Litfin, Karen T. 1994. *Ozone Discourses: Science and Politics in Global Environmental Cooperation.* New York: Columbia University Press.

Litfin, Karen T. 1995. "Framing Science: Precautionary Discourse and the Ozone Treaties." *Millennium* 24, no. 2: 251–77.

Littell, Richard. 1992. *Endangered and Other Protected Species: Federal Law and Regulation.* Washington, D.C.: Bureau of National Affairs.

Lock, Reiner, and Dennis P. Hartwick, eds. 1991. *The New Clean Air Act: Compliance and Opportunity.* Arlington, Va.: Public Utilities Reports.

Lones, Laura L. 1989. "The Marine Mammal Protection Act and International Protection of Cetaceans: A Unilateral Attempt to Effectuate Transnational Conservation." *Vanderbilt Journal of Transnational Law* 22 (4): 997–1028.

Lovejoy, Paul E. 1983. *Transformations in Slavery: A History of Slavery in Africa.* Cambridge: Cambridge University Press.

Lyster, Simon. 1985. *International Wildlife Law.* Cambridge: Grotius Publications.

Main, Jeremy. 1988. "Here Comes the Big New Cleanup." *Fortune,* 118 (21 November): 102–18.

Malloy, Michael P. 1990. *Economic Sanctions and U.S. Trade.* Boston: Little, Brown.

Margavio, Anthony V., and Craig J. Forsyth. 1996. *Caught in the Net: The Conflict between Shrimpers and Conservationists.* College Station, Tex.: Texas A&M University Press.

Marquez, Humberto. 1996. "Shrimp, Next on the List for U.S. Decertification." *Inter Press Service,* 25 April (Lexis/Nexis).

Marshall, Eliot. 1984. "Canada Goes It Alone on Acid Rain Controls." *Science* 226 (21 December): 1275.

Martin, Gene S., Jr., and James W. Brennan. 1989. "Enforcing the International Convention for the Regulation of Whaling: The Pelly and Packwood-Magnuson Amendments." *Denver Journal of International Law and Policy* 17(2): 293–315.

Martin, Lisa L. 1992. *Coercive Cooperation: Explaining Multilateral Economic Sanctions.* Princeton: Princeton University Press.

Mathiesen, Heidi. 1982. "Antarctica: Cutting up a Frozen Pie." *Christian Science Monitor,* 16 April: 22 (Lexis/Nexis).

Matthews, Jessica Tuchman. 1991. "Introduction and Overview." In *Preserving the Global Environment: The Challenge of Shared Leadership,* edited by Jessica Tuchman Matthews, 15–38. New York: W. W. Norton.

M'Gonigle, R. Michael, and Mark W. Zacher. 1979. *Pollution, Politics, and International Law: Tankers at Sea.* Berkeley and Los Angeles: University of California Press.

McDonnell, Patrick. 1986. "U.S. Reportedly Will Lift Embargo on Tuna from Mexico." *Los Angeles Times,* 6 August: 3 (Lexis/Nexis).

McKinney, Michael L., and Robert M. Schoch. 1996. *Environmental Science: Systems and Solutions.* Minneapolis/St. Paul: West.

McLean, William, and Sompong Sucharitkul. 1988. "Fisheries Management and Development in the EEZ: The North, South, and Southwest Pacific Experience." *Notre Dame Law Review* 63(4): 492–534.

"Mexico News Briefs." 1984. *U.P.I.,* 4 March (Lexis/Nexis).

"Mexico Terminating All Fishing Accords with United States." 1980. *New York Times,* 29 December: A1 (Lexis/Nexis).

Meyer, Stephen M. 1992. "Environmentalism Doesn't Steal Jobs." *New York Times,* 26 March: A23.

Miller, Susan Katz. 1993. "Will U.S. Sanctions Save the Rhino?" *New Scientist* 137 (1859): 9 (Lexis/Nexis).

Molloy, Michael P. 1992. *Economic Sanctions and U.S. Trade.* Boston: Little, Brown.

Morgenthau, Hans J. 1967. *Politics among Nations: The Struggle for Power and Peace.* 4th ed. New York: Alfred A. Knopf.

Morrisette, Peter. 1989. "The Evolution of Policy Responses to Stratospheric Ozone Depletion." *Natural Resources Journal* 29: 793–820.

Morrison, Patt. 1986. "U.S. Imposes Wildlife Ban on Singapore." *Los Angeles Times,* 3 October: 1.

Mosley, Paul, Jane Harrigan, and John Toye. 1991. *Aid and Power: The World Bank and Policy-Based Lending.* Vol. 1. London: Routledge.

Mueller, Gene. 1989. "U.S. Authorities Are Too Soft on Foreign Fish Netters." *Washington Times,* 15 August: D8 (Lexis/Nexis).

Murdock, Clark A. 1977. "Economic Factors as Objects of Security: Economics, Security, and Vulnerability." In *Economic Issues and National Security,* edited by Klaus Knorr and Frank N. Trager, 67–98. Lawrence, Kan.: Regents Press of Kansas, for the National Security Education Program.

Nangle, Orval E. 1989. "Stratospheric Ozone: United States Regulation of Chlorofluorocarbons." *Environmental Affairs* 16 (summer): 531–80.

National Research Council. 1981. *Atmosphere-Biosphere Interactions: Toward a Better Understanding of the Consequences of Fossil Fuel Combustion.* Washington, D.C.: National Academy Press.

National Research Council. 1990. *Decline of the Sea Turtles: Causes and Prevention.* Washington, D.C.: National Academy Press.

National Research Council. 1992. *Dolphins and the Tuna Industry.* Washington, D.C.: National Academy Press.

Nordhaus, William D. 1992. "An Optimal Transition Path for Controlling Greenhouse Gases." *Science* 258 (20 November): 1315–19.

Norsworthy, J. R., Michael J. Harper, and Kent Kunze. 1979. "The Slowdown in Productivity Growth: Analysis of Some Contributing Factors." *Brookings Papers on Economic Activity* 2: 387–421.

Noss, Reed F. 1991. "From Endangered Species to Biodiversity." In *Balancing on the Brink of Extinction: The Endangered Species Act and Lessons for the Future,* edited by Kathryn A. Kohn, Washington, D.C.: Island Press.

"Now the Squeeze on Metals." 1979. *Business Week,* 2 July: 46 (Lexis/Nexis).

Nye, Joseph S., Jr. 1990. *Bound to Lead: The Changing Nature of American Power.* New York: Basic Books.

Olson, Mancur. 1965. *The Logic of Collective Action: Public Goods and the Theory of Groups*. Cambridge, Mass.: Harvard University Press.

Olson, Mancur. 1982. *The Rise and Decline of Nations: Economic Growth, Stagflation, and Social Rigidities*. New Haven: Yale University Press.

Organization for Economic Cooperation and Development. 1985. *Environment and Economics*. Paris: OECD.

Organisation for Economic Cooperation and Development. 1985. *Environmental Policy and Technical Change*. Paris: OECD.

Organisation for Economic Cooperation and Development. 1992. *The OECD Environment Industry: Situation, Prospects, and Government Policies*. Paris: OECD.

Organisation for Economic Cooperation and Development. 1993. *Environmental Policies and Industrial Competitiveness*. Paris: OECD.

Oye, Kenneth A., and James H. Maxwell. 1995. "Self-Interest and Environmental Management." *Journal of Theoretical Politics* 6 (4): 192–221.

Paine, C. 1969. "A Note on Trophic Complexity and Community Stability." *American Naturalist* 103: 91.

Paine, R. T. 1966. "Food Web Complexity and Species Diversity." *American Naturalist* 100: 65–75.

"Panama Denies Violating Tuna Conservation Efforts." 1996. *Reuters World Service*, 20 November (Lexis/Nexis).

"Panel Wants Trade Sanctions on Taiwan Times." 1994. *Reuters*, 7 April (Lexis/Nexis).

Parfit, Michael. 1995. "Diminishing Returns: Exploiting the Ocean's Bounty." *National Geographic*, (November): 2–37.

Parrish, Michael, and Juanita Darlin. 1992. "Mexico Backs away from Pact on Tuna." *Los Angeles Times*, 4 November: D2 (Lexis/Nexis).

Parson, Edward A. 1993. "Protecting the Ozone Layer." In *Institutions for the Earth: Sources of Effective International Environmental Protection*. Edited by Peter M. Haas, Robert O. Keohane, and Marc A. Levy. Cambridge, Mass.: MIT Press.

Paskura, Carl A., Jr., and Devorah Vaughn Nestor. 1992. "Environmental Protection Agency, Trade Effects of the 1990 Clean Air Act Amendments." Report.

Pedrozo, Raul. 1993. "The International Dolphin Conservation Act of 1992: Unreasonable Extension of U.S. Jurisdiction in the Eastern Tropical Pacific Ocean Fishery." *Tulane Environmental Law Journal*(7): 77–130.

Peterson, Clifford L., and William H. Bayliff. 1985. "Organization, Functions, and Achievements of the Inter-American Tropical Tuna Commission." *Inter-American Tropical Tuna Commission Special Report No. 5*. La Jolla, Calif.: IATTC.

Peterson, M. J. 1993. "International Fisheries Management." In *Institutions for the Earth: Sources of Effective International Environmental Protection*, edited by Peter M. Haas, Robert O. Keohane, and Marc A. Levy, 249–305. Cambridge, Mass.: MIT Press.

Pleming, Sue. 1993. "U.N. Body Cites China, Taiwan over Rhino Horn." *Reuters,* 7 September (Lexis/Nexis).

"PM Lien Terms US Sanctions 'Unfair' and 'Regretful.'" 1994. *Central News Agency,* 1 April (Lexis/Nexis).

"Porpoise-kill Limit Set by House." 1977. *Facts on File World News Digest* 11 June: 442D3 (Lexis/Nexis).

Portnoy, Paul R. 1981. "The Macroeconomic Impacts of Federal Environmental Regulation." In *Environmental Regulation and the U.S. Economy,* edited by Henry M. Peskin, Paul R. Portnoy, and Allan V. Kneese, 25–54. Baltimore: Johns Hopkins University Press.

Porter, Michael E. 1990. *The Competitive Advantage of Nations.* London: MacMillan Press.

Porter, Michael E. 1991. "America's Green Strategy." *Scientific American,* 269 (April): 168.

Pritchard, Chris. 1984. "Tuna Fishing Causes a Row between U.S. and Pacific Islands." *Christian Science Monitor,* 9 October: 16 (Lexis/Nexis).

Putnam, Robert D. 1988. "Diplomacy and Domestic Politics: The Logic of Two-Level Games." *International Organization* 42 (summer): 427–60.

R.C.C. 1981. "Congress Debates Depletion of Ozone in the Stratosphere." *Christian Science Monitor,* 14 October: 19 (Lexis/Nexis).

Rasmussen, N. Peter. 1981. "The Tuna War: Fishery Jurisdiction in International Law." *University of Illinois Law Review* (3): 755–74.

"Reagan Signs Tuna Law." 1984. *Associated Press,* 4 October (Lexis/Nexis).

"Regional News—Utah." 1991. *U.P.I.,* 8 February (Lexis/Nexis).

Reinhold, Robert. 1981. "As Others Seek to Exploit Antarctic, U.S. Takes the Scientific Approach." *New York Times,* 21 December: D13 (Lexis/Nexis).

Ripley, Randall B., and Grace A. Franklin. 1980. *Congress, the Bureaucracy, and Public Policy, Revised Edition.* Homewood, Ill.: Dorsey Press.

Rosenbaum, Walter A. 1995. *Environmental Politics and Policy.* 3rd ed. Washington, D.C.: CQ Press.

Rosecrantz, Richard. 1986. *The Rise of the Trading State.* New York: Basic Books.

Routt, Lt. J. Allison. 1998. Interview. National Marine Fisheries Service, Southwest Region, 21 October.

Routt, Lt. J. Allison. 1999. Interview. National Marine Fisheries Service, Southwest Region, 29 April.

Rowlands, Ian H. 1995. *The Politics of Global Atmospheric Change.* Manchester: Manchester University Press.

Rudloe, Jack, and Anne Rudloe. 1989. "Shrimpers and Lawmakers Collide over a Move to Save the Sea Turtle." *Smithsonian,* 20 (December), pp. 44ff. (Lexis/Nexis).

Safina, Carl. 1995. "The World's Imperiled Fish." *Scientific American,* 273 (November): 46–53.

Schelling, Thomas C. 1960. *The Strategy of Conflict.* Cambridge, Mass.: Harvard University Press.

Schmitten, Rolland A. 1999. "Fishing 'Green' at Sea." Letter to the editor. *Washington Post,* 30 January: A18.

Schomberg, William. 1993. "Mexico: Prawns in the Export Game." *El Financiero International,* 17 May: 10.

Schreurs, Miranda A., and Elizabeth C. Economy. 1997. *The Internationalization of Environmental Protection.* Cambridge: Cambridge University Press.

"Sea Shepherd's [sic] Claims Victory over Driftnet Fleets." 1987. *U.P.I.,* 21 July (Lexis/Nexis).

Sebenius, James K. 1991. "Crafting a Winning Coalition: Negotiating a Regime to Control Global Warming." In *Greenhouse Warming: Negotiating a Global Regime,* 69–98. Washington, D.C.: World Resources Institute.

"Senior World Wildlife Fund Officials . . ." 1979. *Reuters,* 2 October (Lexis/Nexis).

Shabecoff, Philip. 1989. "Huge Drifting Nets Raise Fears for an Ocean's Fish." *New York Times,* 21 March: C1.

Shaw, Roderick W. 1993. "Acid-Rain Negotiations in North America and Europe: A Study in Contrast." In *International Environmental Negotiations,* edited by Gunner Sjostedt, 84-109. Newbery Park: Sage Publications.

Shimberg, Steven J. 1991. "Stratospheric Ozone and Climate Protection: Domestic Legislation and the International Process." *Environmental Law* 21(summer): 2175–2216.

Sjostedt, Gunner, and Bertran I. Spector. 1993. "Conclusion." In *International Environmental Negotiation,* edited by Gunner Sjostedt, 291–314. Newbury Park: Sage Publications.

Snidal, Duncan. 1985. "Coordination versus Prisoner's Dilemma: Implications for International Cooperation and Regimes." *American Political Science Review* 79: 923–42.

Socolow, R., C. Andrews, F. Berkhout, and V. Thomas. 1994. *Industrial Ecology and Global Change.* Cambridge: Cambridge University Press.

Sprinz, Detlef, and Tapani Vaahtoranta. 1994. "The Interest-Based Explanation of International Environmental Policy." *International Organization* 48 (winter): 77–105.

Stanfield, Rochelle L. 1985. "Environmentalists Try the Backdoor Approach to Tackling Acid Rain." *National Journal* 17(2): 2365ff. (Lexis/Nexis).

Stein, Arthur A. 1982. "Coordination and Collaboration: Regimes in an Anarchic World." *International Organization* 36: 299–324.

Stewart, Richard B. 1993. "Environmental Regulation and International Competitiveness." *Yale Law Journal* 102: 2039–2106.

Stewart, Robert. 1992. "Stratospheric Ozone Protection: Changes Over Two Decades of Regulation." *Natural Resources and the Environment* 7(2)(fall): 24–27, 53–54.

Sumi, Kazuo. 1989. "The 'Whale War' between Japan and the United States: Problems and Prospects." *Denver Journal of International Law and Policy* 17(2): 317–72.

Sun, Marjorie. 1984. "Lawsuit Seeks a Cap on Fluorocarbon Production." *Science* 226 (14 December): 1297.

Szekely, Alberto. 1989. "Yellow-Fin Tuna: A Transboundary Resource of the Eastern Pacific." *Natural Resources Journal* 29(fall): 1051–65.

Szekely, Alberto, and Barbara Kwiatkowska. 1992. "Marine Living Resources." In *The Effectiveness of International Environmental Agreements: A Survey of Existing Legal Instruments,* edited by Peter H. Sand, 256–301. Cambridge, UK: Grotius Publications.

"Taiwan May Sue U.S." 1991. *Journal of Commerce,* 4 December: 5A.

"Taiwan Regrets US Trade Sanction Decision." 1994. *Agence France Presse,* 12 April (Lexis/Nexis).

Tedrick, Dan. 1977. "American Tuna Fleet . . ." *Associated Press,* 3 May (Lexis/Nexis).

Tighe, Michael. 1996. "Tuna Commission Approves Sanctions against Three Countries." *AP Worldstream,* 29 November (Lexis/Nexis).

Tobey, James A. 1990. "The Effects of Domestic Environmental Policies on Patterns of World Trade: An Empirical Test." *Kyklos,* 43: 191–209.

Tobey, James A. 1993. "The Impact of Domestic Environmental Policies on International Trade." In *Environmental Policies and Industrial Competitiveness.* Paris: OECD.

Tonelson, Alan, and Lori Wallach. 1996. "We Told You So: The WTO's First Trade Decision Vindicates the Warnings of Critics." *Washington Post,* 5 May: C4 (Lexis/Nexis).

"The Tragedy of the Oceans." 1994. *Economist,* 19 March: 21–24.

"Tuna Boats Back at Sea." 1979. *Washington Post,* 16 November: A16.

"Tuna Fishing: Tuna Accord Approval Not Universal." 1995. *Europe Environment,* 14 November (Lexis/Nexis).

"Tuna War Escalates between U.S. and Ecuador." 1980. *Associated Press* (International News), 17 November (Lexis/Nexis).

"Turtles: Trade in SE Asia Raises Threat of Extinction." 1999. *Greenwire,* 4 May.

United Nations. 1973. *Commodity Trade Statistics.* New York: United Nations Publications.

United Nations. 1977. *Commodity Trade Statistics.* New York: United Nations Publications.

United Nations. 1979. *Commodity Trade Statistics.* New York: United Nations Publications.

United Nations. 1984. *Commodity Trade Statistics.* New York: United Nations Publications.

United Nations. 1985. *Commodity Trade Statistics.* New York: United Nations Publications.

United Nations. 1986. *Commodity Trade Statistics.* New York: United Nations Publications.

United Nations. 1987. *Commodity Trade Statistics.* New York: United Nations Publications.

United Nations. 1989. *Commodity Trade Statistics.* New York: United Nations Publications.

United Nations. 1991. *Commodity Trade Statistics.* New York: United Nations Publications.

United Nations. 1992. *Commodity Trade Statistics.* New York: United Nations Publications.

United Nations. 1993. *Commodity Trade Statistics.* New York: United Nations Publications.

United Nations. 1994. *Commodity Trade Statistics.* New York: United Nations Publications.

United Nations. 1995. *Commodity Trade Statistics.* New York: United Nations Publications.

United Nations. 1996. *Commodity Trade Statistics.* New York: United Nations Publications.

United Nations. 1997. *Commodity Trade Statistics.* New York: United Nations Publications.

United Nations General Assembly Resolution 44/225.

*UPI Wire Report.* 1994. 11 April (Lexis/Nexis).

"U.S. Considers Trade Sanctions against Taiwan, South Korea." 1991. *Journal of Commerce,* 15 August: 3A (Lexis/Nexis).

"U.S. Curbs on Shrimp Import Not to Affect India." 1999. *The Hindu,* FT Asia Intelligence Wire, 6 May (Lexis/Nexis).

"U.S. May Ban Shrimp Imports from Latin American Nations." 1991. *Journal of Commerce,* 9 January: 4A (Lexis/Nexis).

"U.S. Might Lift Ban on Venezuelan Tuna." 1988. *Journal of Commerce,* 19 October: 4A (Lexis/Nexis).

"U.S. Ordered to Curb Japan on Whaling." 1985. *Los Angeles Times,* 6 August: 2 (Lexis/Nexis).

"U.S.: Seven Countries Sign Dolphin Protection Agreement." 1998. *AA Newsfeed,* 22 May (Lexis/Nexis).

"U.S. Shrimp Ban Seen Costly for Trinidadians." 1995. *Journal of Commerce,* 1 June: 5A (Lexis/Nexis).

"U.S. Tells Shrimpers to Give Sea Turtles an Escape 'Door.'" 1989. *New York Times,* 6 September: B8 (Lexis/Nexis).

Van Dyke, Jon, and Carolyn Nicol. 1987. "U.S. Tuna Policy: A Reluctant Acceptance of the International Norm." In *Tuna Issues and Perspectives in the Pacific Islands Region,* edited by David J. Doulman, 105–132. Honolulu, Hawaii: East-West Center.

Vaughan, Ray. 1994. *Endangered Species Act Handbook.* Rockville, Md.: Government Institutes.

Vogel, David. 1995. *Trading Up: Consumer and Environmental Regulation in a Global Economy.* Cambridge, Mass.: Harvard University Press.

Waltz, Kenneth. 1979. *Theory of International Politics.* New York: Random House.

Wang, James C. F. 1992. *Handbook on Ocean Politics and Law.* New York: Greenwood Press.

Weber, Max. 1978. *Economy and Society: An Outline of Interpretive Sociology.* Berkeley and Los Angeles: University of California Press.

Weintraub, Sidney, ed. 1982. *Economic Coercion and U.S. Foreign Policy: Implications of Case Studies from the Johnson Administration.* Boulder: Westview Press.

Wenner, Lettie. 1993. "Transboundary Problems in International Law." In *Environmental Politics in the International Arena,* edited by Sheldon Kamieniecki, 165–178. Albany: State University Press of New York.

Williams, Frances. 1998. "U.S. Appeal on Shrimp Import Ban Rejected." *Financial Times,* 13 October: 10.

Williams, Ted. 1992. "The Last Bluefin Hunt." *Audubon,* (July-August): 14–20.

Woodyard, Chris. "Shrimp Caught in Restrictions, U.S. Squeeze on Imports May Up Prices." 1996. *Houston Chronicle,* 2 November: 1 (Lexis/Nexis).

World Bank. 1992. *World Development Report 1992: Development and the Environment.* Oxford: Oxford University Press.

"World Briefs: Group Approves Ban on Fishing with Driftnets." 1998. *St. Louis Post-Dispatch,* 9 June: A5 (Lexis/Nexis).

World Conservation Monitoring Center. 1996. "Global Top Twenty." http://wcmc.org.uk/infoserve/species/sp_top20.html, date visited: 11 June.

World Meterological Organization. 1998. "Summary of Scientific Assessment of the Ozone Layer 1998." World Meteorological Organization Press Release, 22 June 1998. http://www.wmo.ch/web/Press/wmo-unep.html, date visited: 2 May 1999.

World Trade Organization. 1998. "United States—Import Prohibition of Certain Shrimp and Shrimp Products." Interim Panel Report, unpublished copy obtained from the Office of the United States Trade Representative.

World Trade Organization. 1998. "United States—Import Prohibition of Certain Shrimp and Shrimp Products." Report of the Appellate Body, AB-1998-4, 12 October.

World Wildlife Fund. 1989. "A Program to Save the African Elephant." *World Wildlife Fund Letter,* no. 2: 1–2.

World Wildlife Fund. 1994. "WWF's 1994 Ten Most Endangered List." http://envirolink.org/arrs/endangered.html, date visited: 9 March 1996.

World Wildlife Fund. 1996. "The Large Pelagic Fishes," http://www.panda.org/research/fishfile2/fish42.htm, date visited: 26 June.

Wu, Sofia. 1994. "Taiwan to Meet CITES Requirements by September." *Central News Agency,* 21 April (Lexis/Nexis).

Yaffee, Steven Lewis. 1982. *Prohibitive Policy: Implementing the Federal Endangered Species Act .* Cambridge, Mass.: MIT Press.

Yandle, Bruce. 1984. "Intertwined Interests, Rent Seeking, and Regulation." *Social Science Quarterly* 1984: 1002–12.

Yandle, Bruce. 1989. *The Political Limits of Environmental Regulation: Tracking the Unicorn.* New York: Quorum Books.

Yohe, Gary W. 1979. "The Backward Incidence of Pollution Control—Some Comparative Statics in General Equilibrium." *Journal of Environmental Economics and Management* 6: 187–98.

Zoller, Elisabeth. 1985. *Enforcing International Law through U.S. Legislation.* Dobbs Ferry, N.Y.: Transnational Publishers.

## United States Government Documents

*Congressional Digest.* 1990. Testimony of Richard L. Lawson, President, National Coal Association, 3 October 1989 in hearings before the Environmental Protection Subcommittee of the Senate Environment and Public Works Committee. 69, no. 3 (March).

*Congressional Record.* 1989. Senate, 101st Congress, 1st Sess., 29 September: s12266, Statements of Mr. Johnson and Mr. Breux.

*Earth Island Institute et al. v Mosbacher et al.* 1990. United States District Court for the Northern District of California, No. C-88-1380-TEH, 746 F. Supp. 964, 28 August.

*Earth Island Institute et al. v Mosbacher et al.* 1991. United States Court of Appeals for the Ninth Circuit, No. 90-16581, 929 F. 2d 1449, 11 April.

*Earth Island Institute et al. v Mosbacher et al.* 1992. United States District Court for the Northern District of California, No. C-88-1380-TEH, 785 F. Supp. 826, 3 February.

*Earth Island Institute et al. v Warren Christopher et al.* 1995. United States Court of International Trade, Court No. 94-06-00321, 913 F. Supp. 599.

*Japan Whaling Association v American Cetacean Society.* 1986. 478 U.S. 221, 105 S. Ct. 2860 (1986).

"Legislative History." 1971. From House Report (Merchant Marine and Fisheries Committee) No. 92-707, 4 December: 4144.

NOAA, NMFS. 1998. "Tuna/Dolphin Embargo Status Update." http://swr.ucsd.edu/psd/embargo2.htm, 1 June, date visited: 13 October.

NOAA, NMFS. 1999. "NOAA Fisheries Headquarters." http://www.nmfs.org, date visited: 21 April.

*U.S. Code Congressional and Administrative News.* 1971: 2409, from House Report 92-468.

*U.S. Code Congressional and Administrative News.* 1971: 3612–3, from Senate Report 2094, 18 July 1950.

*U.S. Code Congressional and Administrative News.* 1990: 6282.

U.S. Congress. Senate. 91st Cong., 2nd sess. Senate Report 91-1196, 1970.

U.S. Congress. House. 92nd Cong., 1st sess. House Report 92-468.

U.S. Congress. Senate. 93rd Cong., 1st sess. Senate, Executive Report 93-14.

U.S. Congress. House. 95th Cong., 2nd sess. House Report 95-1029.

U.S. Congress. Senate. Subcommittees on Environmental Protection and Hazardous Wastes and Toxic Substances of the Committee on Environment and Public Works. *Ozone Depletion, the Greenhouse Effect, and Climate Change.* Pt. 2. 100th Cong., 1st sess., 28 January 1987.

U.S. Congress. Senate. Subcommittees on Environmental Protection and Hazardous Wastes and Toxic Substances of the Committee on Environment and Public Works. *Stratospheric Ozone Depletion and Chlorofluorocarbons.* 100th Cong., 1st sess., 12, 13, and 14 May, 1987.

U.S. Congress. House. Subcommittee on Fisheries and Wildlife Conservation and the Environment of the Committee on Merchant Marine and Fisheries. "Sea Turtle Conservation and the Shrimp Industry." 101st Cong., 2nd sess., Serial No. 101-83, 1 May 1990.

U.S. Congress. House. 101st Cong., 1st sess. "Cooperative Measures on Driftnet Fishing: Communication from the President of the United States." House Doc. 101-93, 7 September 1989.

U.S. Congress. House. 1st sess. "Sea Turtle Activities in Japan: Message from the President of the United States Transmitting a Report on Certification by the Secretaries of the Interior and Commerce Concerning Activities by Nationals of Japan Engaging in Trade in Sea Turtles That Threatens the Survival of Two Endangered Species and Severely Diminishes the Effectiveness of the Convention on International Trade, pursuant to 22 U.S.C. 1978(a)(2)." House Document 102-85, 20 May 1991.

U.S. Congress. House. 102nd Cong., 1st sess. "Driftnet Fishing Violations of the Republic of Korea and Taiwan: Message from the President of the United States," House Document 102-155, 21 October 1991.

U.S. Congress. House. 103rd Cong., 1st sess., "Violations Relating to Endangered Species: Message from the President of the United States Transmitting a Report Concerning the People's Republic of China and Taiwan Engaging in Trade of Rhinoceros and Tiger Parts and Products that Diminishes the Effectiveness of the Convention on International Trade in Endangered Species of Wild Fauna and Flora (CITES), pursuant to 22 U.S.C. 1978 (b)." House Document 103–162, 8 November 1993.

United States Congress. House, 95th Cong., 1st sess. 143 *Congressional Record*, 21 January 1997: H242.

United States Department of State Press Statement. 1999. "Sea Turtle Conservation and Shrimp Imports." 4 May.

## Code of Federal Regulations

40 C.F.R. 60

50 C.F.R. 17.11

50 C.F.R. 18.13

50 C.F.R. 23.23

50 C.F.R. 281.7

## Federal Register

39 Federal Register 2483

39 Federal Register 32118–19

39 Federal Register 32124

40 Federal Register 819

40 Federal Register 8239

40 Federal Register 48159

40 Federal Register 42230

40 Federal Register 48160

40 Federal Register 56904

42 Federal Register 54294

42 Federal Register 56617

42 Federal Register 64121

42 Federal Register 64548

42 Federal Register 64558

42 Federal Register 64559–60

43 Federal Register 1693-94

43 Federal Register 3566

43 Federal Register 5521

43 Federal Register 11301

43 Federal Register 11318

43 Federal Register 31144–45

43 Federal Register 36263

43 Federal Register 40025

44 Federal Register 10172

44 Federal Register 25554

45 Federal Register 47562
44 Federal Register 47431
44 Federal Register 57100–101
44 Federal Register 53118
44 Federal Register 59985
45 Federal Register 7363
45 Federal Register 9284–85
45 Federal Register 11971
45 Federal Register 13094–95
45 Federal Register 58459
45 Federal Register 75215–16
45 Federal Register 77219
46 Federal Register 10974
47 Federal Register 8446
47 Federal Register 11307–8
47 Federal Register 15151
48 Federal Register 14431
48 Federal Register 16798
48 Federal Register 30422
48 Federal Register 32832
48 Federal Register 56986
49 Federal Register 33526
50 Federal Register 15273
50 Federal Register 30950–51
51 Federal Register 15571
51 Federal Register 28963
51 Federal Register 29183
51 Federal Register 34159
51 Federal Register 36504
51 Federal Register 36864
51 Federal Register 47064
52 Federal Register 24244–62
53 Federal Register 8910
53 Federal Register 39743
53 Federal Register 45953
54 Federal Register 50763

55 Federal Register 42236

55 Federal Register 48666

56 Federal Register 1051

56 Federal Register 12367

56 Federal Register 26995

56 Federal Register 30379

57 Federal Register 18446ff

57 Federal Register 2170

57 Federal Register 17857

57 Federal Register 17858

57 Federal Register 38549

57 Federal Register 47620–24

57 Federal Register 59979

57 Federal Register 61597

58 Federal Register 3013–14

58 Federal Register 9015–7

58 Federal Register 30082

58 Federal Register 40685

58 Federal Register 42030

59 Federal Register 15655

59 Federal Register 22043–45

59 Federal Register 25697

59 Federal Register 35911

59 Federal Register 65974

60 Federal Register 10332

60 Federal Register 24962

60 Federal Register 43640–41

61 Federal Register 17342–44

61 Federal Register 18721–22

61 Federal Register 24998–99

61 Federal Register 43395

61 Federal Register 59482

62 Federal Register 4826

62 Federal Register 19157

62 Federal Register 23479–80

62 Federal Register 29759

62 Federal Register 44422–23
63 Federal Register 30550–51
63 Federal Register 44499

## Public Laws
P.L. 84-159; 69 Stat. 322 (1955)
P.L. 88-206; 77 Stat. 392 (1963)
P.L. 89-206; 79 Stat. 992 (1965)
P.L. 90-148; 81 Stat. 485 (1967)
P.L. 95-95; 91 Stat. 726 (1977)
P.L. 88-206; 77 Stat. 392 (1963)
P.L. 89-669; 80 Sat. 926 (1966)
P.L. 91-135; 83 Stat. 275 (1969)
P.L. 91-135, Section 2
P.L. 91-135, Section 5
P.L. 91-604; 84 Stat. 1676 (1970)
P.L. 92-219, sec. 8, (g)(3) and (4)
P.L. 92-522
P.L. 93-205; 87 Stat. 884 (1973)
P.L. 95-376
P.L. 100-220, 101 Stat. 1477-8
P.L. 100-478 Sec. 2201(b)(1). 102 Stat. 2318
P.L. 100-478 Sec. 2202(a)(1) Sec. 2202(2); 102 Stat. 2319
P.L. 100-478, Title II, §2303, Oct. 7, 1988, 102 Stat. 2322
P.L. 100-711
P.L. 101-162, §609, (b) (2)
P.L. 101-549; 104 Stat. 2399 (1990)
P.L. 101-627, Title I, Section 101(b), 107; 104 Stat. 4443
P.L. 102-582; 106 Stat. 4900, Section 101
P.L. 101-549, 104 Stat. 2399
P.L. 104-42, 111 Stat. 1122-1139

## United States Code
16 U.S.C. 182
16 U.S.C. 1361-1407
16 U.S.C. 1532
16 U.S.C. 1533

16 U.S.C. 1822
16 U.S.C. 1825
16 U.S.C. 2435
16 U.S.C. 3372
16 U.S.C. 4242
22 U.S.C. 1978
22 U.S.C. 1974
42 U.S.C. 7457
42 U.S.C. 7502

# Index